Civil Engineering: Design, Construction and Maintenance of Buildings

Civil Engineering: Design, Construction and Maintenance of Buildings

Amanda Wang

Larsen & Keller
www.larsen-keller.com

Civil Engineering: Design, Construction and Maintenance of Buildings
Amanda Wang
ISBN: 978-1-64172-395-4 (Hardback)

Larsen & Keller
Published by Larsen and Keller Education,
5 Penn Plaza,
19th Floor,
New York, NY 10001, USA

Cataloging-in-Publication Data

Civil engineering : design, construction and maintenance of buildings / Amanda Wang.
p. cm.
Includes bibliographical references and index.
ISBN 978-1-64172-395-4
1. Civil engineering. 2. Architecture--Designs and plans. 3. Buildings--Maintenance.
4. Buildings--Specifications. 5. Building. I. Wang, Amanda.
TA145 .C58 2020
624--dc23

Printed and bound in China.

For more information regarding Larsen and Keller Education and its products, please visit the publisher's website www.larsen-keller.com

Table of Contents

Preface

This book has been written, keeping in view that students want more practical information. Thus, my aim has been to make it as comprehensive as possible for the readers. I would like to extend my thanks to my family and co-workers for their knowledge, support and encouragement all along.

Civil engineering focuses on the design, maintenance and construction of the physical and naturally built environment. It is a branch of engineering that deals with public works such as roads, bridges, dams, canals, sewerage systems, pipelines, airports, structural components of buildings and railways. Some of the sub-disciplines of civil engineering are coastal engineering, earthquake engineering, forensic engineering, geotechnical engineering, structural engineering, construction engineering and environmental engineering. Coastal engineering deals with the specific demands which arise during the construction at or near the coast, along with the development of the coast itself. Earthquake engineering is concerned with the designing of structures which are expected to withstand hazardous earthquake exposures. This book outlines the processes and applications of civil engineering in detail. Some of the diverse topics covered herein address the varied branches that fall under this category. For someone with an interest and eye for detail, this book covers the most significant topics in this field.

A brief description of the chapters is provided below for further understanding:

Chapter - Introduction

Civil engineering is a discipline that is concerned with the construction, design and maintenance of the physical and naturally built environment. One of its important branches is foundation engineering. This is an introductory chapter which will introduce briefly all the significant aspects of civil engineering.

Chapter - Construction and Building Engineering

The discipline which deals with the management, construction, planning and designing of infrastructures like airports, railroads, bridges, dams, utilities, tunnels, roads, etc. is known as construction engineering. This chapter closely examines the key concepts related to construction engineering to provide an extensive understanding of the subject.

Chapter - Structural Engineering and Design

Structural engineering is a sub-discipline of civil engineering that focuses on the framework of structures and ensures that they are designed in a way to withstand the stresses and pressures of their environment. It studies various structural materials such as steel frames, reinforced concrete, prestressed concrete, etc. The topics elaborated in this chapter will help in gaining a better perspective about these focus areas of structural engineering.

Chapter - Understanding Geotechnical Engineering

Geotechnical engineering refers to the branch of civil engineering that deals with the application of engineering principles for the acquisition, interpretation and usage of knowledge about the materials of the Earth's crust for solving engineering problems. It uses principles of various fields such as soil mechanics, soil compaction, soil gradation, erosion control, soil nailing, etc. This chapter discusses in detail these principles and theories related to geotechnical engineering.

Chapter – Earthquake Engineering

The interdisciplinary branch of engineering that analyzes and designs structures like buildings and bridges with regard to earthquakes is known as earthquake engineering. It's used to design earthquake-resistant structures and implement active vibration control. This chapter has been carefully written to provide an easy understanding of these applications of earth-quake engineering.

Chapter – Hydraulic Engineering: A Comprehensive Study

Hydraulic engineering deals with the application of fluid mechanics for solving problems related to the collection, storage, control and use of water. It also makes use of principles from the fields of fluid dynamics and hydrostatics. All these diverse aspects of hydraulic engineering have been carefully analyzed in this chapter.

Amanda Wang

1
Introduction

Civil engineering is a discipline that is concerned with the construction, design and maintenance of the physical and naturally built environment. One of its important branches is foundation engineering. This is an introductory chapter which will introduce briefly all the significant aspects of civil engineering.

Civil Engineering

Civil engineering is the field concerned with the designing and executing structural works that serve the general public. The term was first used in the 18th century to distinguish the newly recognized profession from military engineering, until then preeminent. From earliest times, however, engineers have engaged in peaceful activities, and many of the civil engineering works of ancient and medieval times—such as the Roman public baths, roads, bridges, and aqueducts; the Flemish canals; the Dutch sea defenses; the French Gothic cathedrals; and many other monuments—reveal a history of inventive genius and persistent experimentation.

Civil Engineering Functions

The functions of the civil engineer can be divided into three categories: those performed before construction (feasibility studies, site investigations, and design), those performed during construction (dealing with clients, consulting engineers, and contractors), and those performed after construction (maintenance and research).

Feasibility Studies

No major project today is started without an extensive study of the objective and without preliminary studies of possible plans leading to a recommended scheme, perhaps with alternatives. Feasibility studies may cover alternative methods—e.g., bridge versus tunnel, in the case of a water crossing—or, once the method is decided, the choice of route. Both economic and engineering problems must be considered.

Site Investigations

A preliminary site investigation is part of the feasibility study, but once a plan has been adopted a more extensive investigation is usually imperative. Money spent in a rigorous study of ground and substructure may save large sums later in remedial works or in changes made necessary in constructional methods.

Since the load-bearing qualities and stability of the ground are such important factors in any large-scale construction, it is surprising that a serious study of soil mechanics did not develop until the mid-1930s. Karl von Terzaghi, the chief founder of the science, gives the date of its birth as 1936, when the First International Conference on Soil Mechanics and Foundation Engineering was held at Harvard University and an international society was formed. Today there are specialist societies and journals in many countries, and most universities that have a civil engineering faculty have courses in soil mechanics.

Design

The design of engineering works may require the application of design theory from many fields—e.g., hydraulics, thermodynamics, or nuclear physics. Research in structural analysis and the technology of materials has opened the way for more rational designs, new design concepts, and greater economy of materials. The theory of structures and the study of materials have advanced together as more and more refined stress analysis of structures and systematic testing has been done. Modern designers not only have advanced theories and readily available design data, but structural designs can now be rigorously analyzed by computers.

Architectural Engineering

Construction

The promotion of civil engineering works may be initiated by a private client, but most work is undertaken for large corporations, government authorities, and public boards and authorities. Many of these have their own engineering staffs, but for large specialized projects it is usual to employ consulting engineers.

The consulting engineer may be required first to undertake feasibility studies, then to recommend a scheme and quote an approximate cost. The engineer is responsible for the design of the works, supplying specifications, drawings, and legal documents in sufficient detail to seek competitive tender prices. The engineer must compare quotations and recommend acceptance of one of them. Although he is not a party to the contract, the engineer's duties are defined in it; the staff must supervise the construction and the engineer must certify completion of the work. Actions must be consistent with duty to the client; the professional organizations exercise disciplinary control over professional conduct. The consulting engineer's senior representative on the site is the resident engineer.

A phenomenon of recent years has been the turnkey or package contract, in which the contractor undertakes to finance, design, specify, construct, and commission a project in its entirety. In this case, the consulting engineer is engaged by the contractor rather than by the client.

The contractor is usually an incorporated company, which secures the contract on the basis of the consulting engineer's specification and general drawings. The consulting engineer must agree to any variations introduced and must approve the detailed drawings.

Maintenance

The contractor maintains the works to the satisfaction of the consulting engineer. Responsibility for maintenance extends to ancillary and temporary works where these form part of the overall construction. After construction a period of maintenance is undertaken by the contractor, and the payment of the final installment of the contract price is held back until released by the consulting engineer. Central and local government engineering and public works departments are concerned primarily with maintenance, for which they employ direct labour.

Branches of Civil Engineering

In 1828 Thomas Tredgold of England wrote:

> The most important object of Civil Engineering is to improve the means of production and of traffic in states, both for external and internal trade. It is applied in the construction and management of roads, bridges, railroads, aqueducts, canals, river navigation, docks and storehouses, for the convenience of internal intercourse and exchange; and in the construction of ports, harbours, moles, breakwaters and lighthouses; and in the navigation by artificial power for the purposes of commerce.

It is applied to the protection of property where natural powers are the sources of injury, as by embankments for the defence of tracts of country from the encroachments of the sea, or the overflowing of rivers; it also directs the means of applying streams and rivers to use, either as powers to work machines, or as supplies for the use of cities and towns, or for irrigation; as well as the means of removing noxious accumulations, as by the drainage of towns and districts to secure the public health.

A modern description would include the production and distribution of energy, the development of aircraft and airports, the construction of chemical process plants and nuclear power stations, and water desalination. These aspects of civil engineering may be considered under the following headings: construction, transportation, maritime and hydraulic engineering, power, and public health.

Construction

Almost all civil engineering contracts include some element of construction work. The development of steel and concrete as building materials had the effect of placing design more in the hands of the civil engineer than the architect. The engineer's analysis of a building problem, based on function and economics, determines the building's structural design.

Transportation

Roman roads and bridges were products of military engineering, but the pavements of McAdam and the bridges of Perronet were the work of the civil engineer. So were the canals of the 18th century and the railways of the 19th, which, by providing bulk transport with speed and economy, lent a powerful impetus to the Industrial Revolution. The civil engineer today is concerned with an even larger transportation field—e.g., traffic studies, design of systems for road, rail, and air, and construction including pavements, embankments, bridges, and tunnels.

Maritime and Hydraulic Engineering

Harbour construction and shipbuilding are ancient arts. For many developing countries today the establishment of a large, efficient harbour is an early imperative, to serve as the inlet for industrial plant and needed raw materials and the outlet for finished goods. In developed countries the expansion of world trade, the use of larger ships, and the increase in total tonnage call for more rapid and efficient handling. Deeper berths and alongside-handling equipment (for example, for ore) and navigation improvements are the responsibility of the civil engineer.

The development of water supplies was a feature of the earliest civilizations, and the demand for water continues to rise today. In developed countries the demand is for industrial and domestic consumption, but in many parts of the world—e.g., the Indus basin—vast schemes are under construction, mainly for irrigation to help satisfy the food demand, and are often combined with hydroelectric power generation to promote industrial development.

Power

Civil engineers have always played an important part in mining for coal and metals; the driving of tunnels is a task common to many branches of civil engineering. In the 20th century the design and construction of power stations has advanced with the rapid rise in demand for electric power, and nuclear power stations have added a whole new field of design and construction, involving prestressed concrete pressure vessels for the reactor.

The exploitation of oil fields and the discoveries of natural gas in significant quantities have initiated a radical change in gas production. Shipment in liquid form from the Sahara and piping from the bed of the North Sea have been among the novel developments.

Dams today are among the largest construction works, and design development is promoted by bodies like the International Commission on Large Dams. The design of large impounding dams in places with population centres close by requires the utmost in safety engineering, with emphasis on soil mechanics and stress analysis. Most governments exercise statutory control of engineers qualified to design and inspect dams.

Public Health

Drainage and liquid-waste disposal are closely associated with antipollution measures and the re-use of water. The urban development of parts of water catchment areas can alter the nature of runoff, and the training and regulation of rivers produce changes in the pattern of events, resulting in floods and the need for flood prevention and control.

Modern civilization has created problems of solid-waste disposal, from the manufacture of durable goods, such as automobiles and refrigerators, produced in large numbers with a limited life, to the small package, previously disposable, now often indestructible. The civil engineer plays an important role in the preservation of the environment, principally through design of works to enhance rather than to damage or pollute.

Foundations and Foundation Engineering

The foundations of the building transfer the weight of the building to the ground. While 'foundation' is a general word, normally, every building has a number of individual foundations. Most buildings have some kind of foundation structure directly below every major column, so as to transfer the column loads directly to the ground.

Since the weight of the building rests on the soil (or rock), engineers have to study the properties of the soil very carefully to ensure that it can carry the loads imposed by the building. It is common for engineers to determine the safe bearing capacity of the soil after such study. As the name suggests, this is the amount of weight per unit area the soil can bear. For example, the safe bearing capacity (SBC) at a location could be 20 T/m_2, or tonnes per square metre. This figure is the maximum the soil can bear, so an engineer will take pains to see that her design does not exceed this figure in any part of the building.

This capacity also changes at different depths of soil. In general, the deeper one digs, the greater the SBC, unless there are pockets of weak soil in the earth. To properly support a building, the soil must be very firm and strong. It is common for the soil near the surface of the earth to be loose and weak. If a building is rested on this soil, it will sink into the earth like a ship in water. Building contractors will usually dig until they reach very firm, strong, soil that cannot be dug up easily before constructing a foundation.

To study the properties of the soil before designing foundations, engineers will ask for a soil investigation to be done. A soil investigation engineer will drill a 4" or 6" hollow pipe into the ground, and will remove samples of the earth while doing so. He will then send these samples to a lab to find out the detailed properties of the soil at every depth. Soil is usually composed of strata, or different layers, each with its own set of properties. Drilling technology today makes it easy and economical to drill to great depths, easily several hundred metres or more, even in hard rock.

The soil investigation team will then prepare a soil investigation report that lists the engineering properties of the soil at regular intervals, say every 2 meters. Based on this deport, engineers designing the structure can decide at what depth of soil to provide the foundations, the type of foundations they should provide, and the size of the foundations.

Every once in a while, engineers will find fill at a site. This occurs when humans have previously dug up the earth there, and then filled it back in. This happens if a quarry was dug or a building built there previously. Since fill is loose and soft and cannot support weight, engineers will dig to a depth below that of the fill, where strong soil is found, and construct foundations there.

The study of soil, and its properties and behavior, is called soil mechanics.

Once the foundations have been built, the loose soil that has been excavated must be put back over and around the foundations. This is called backfilling. Backfilling must be done carefully, as the soil there must support the weight of the floor slab at ground level. Backfilling is done by putting back the soil in horizontal layers about a foot thick, and then compacting the earth, or squeezing it under pressure in a wet condition. This causes the soil particles to be squished together and removes air voids, there by making the layers strong. Good backfilling also improves the performance of the foundations, as the earth holds them firmly in place, and weighs down on the foundations to anchor them in position.

The act of strengthening a foundation is called underpinning. It can also be called foundation repair. While it is difficult to conceive of how foundations (that are underground in an occupied building) can be repaired or strengthened, there are techniques that can achieve this. In a sense, these methods can be likened to surgery performed on a patient.

Types of Foundations

all foundations are divided into two categories: shallow foundations and deep foundations. The words shallow and deep refer to the depth of soil in which the foundation is made. Shallow foundations can be made in depths of as little as 3ft (1m), while deep foundations can be made at depths of 60 - 200ft (20 - 65m). Shallow foundations are used for small, light buildings, while deep ones are for large, heavy buildings.

Shallow Foundations

Shallow foundations are also called spread footings or open footings. The 'open' refers to the fact that the foundations are made by first excavating all the earth till the bottom of the footing, and then constructing the footing. During the early stages of work, the entire footing is visible to the eye, and is therefore called an open foundation. The idea is that each footing takes the concentrated load of the column and spreads it out over a large area, so that the actual weight on the soil does not exceed the safe bearing capacity of the soil.

There are several kinds of shallow footings: individual footings, strip footings and raft foundations.

In cold climates, shallow foundations must be protected from freezing. This is because water in the soil around the foundation can freeze and expand, thereby damaging the foundation. These foundations should be built below the frost line, which is the level in the ground above which freezing occurs. If they cannot be built below the frost line, they should be protected by insulation: normally a little heat from the building will permeate into the soil and prevent freezing.

Individual Footings

Individual footings awaiting concreting of the footing column.

Individual footings are one of the most simple and common types of foundations. These are used when the load of the building is carried by columns. Usually, each column will have its own footing. The footing is just a square or rectangular pad of concrete on which the column sits. To get a very rough idea of the size of the footing, the engineer will take the total load on the column and divide it by the safe bearing capacity (SBC) of the soil. For example, if a column has a vertical load of 10T, and the SBC of the soil is $10T/m^2$, then the area of the footing will be $1m^2$. In practice, the designer will look at many other factors before preparing a construction design for the footing.

Individual footings connected by a plinth beam. Note that the footings have been cast on top of beds of plain cement concrete (PCC), which has been done to create a level, firm base for the footing.

Individual footings are usually connected by a plinth beam, a horizontal beam that is built at ground or below ground level.

Strip Footings

Strip footings are commonly found in load-bearing masonry construction, and act as a long strip that supports the weight of an entire wall. These are used where the building loads are carried by entire walls rather than isolated columns, such as in older buildings made of masonry.

Raft or Mat Foundations

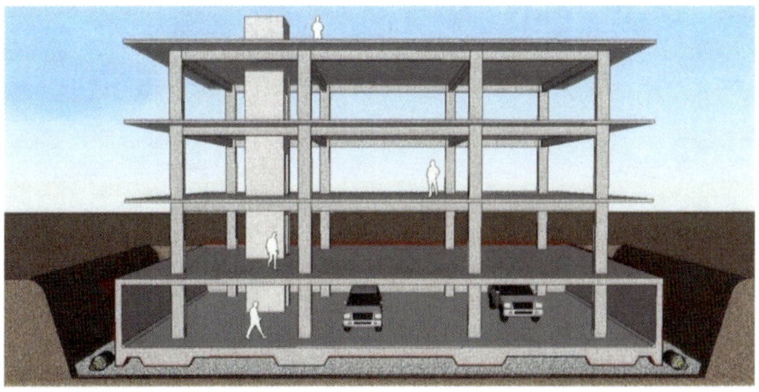

Raft Foundations, also called Mat Foundations, are most often used when basements are to be constructed. In a raft, the entire basement floor slab acts as the foundation; the weight of the building is spread evenly over the entire footprint of the building. It is called a raft because the building is like a vessel that 'floats' in a sea of soil.

Mat Foundations are used where the soil is week, and therefore building loads have to be spread over a large area, or where columns are closely spaced, which means that if individual footings were used, they would touch each other.

Deep Foundations

Pile Foundations

A pile is basically a long cylinder of a strong material such as concrete that is pushed into the ground so that structures can be supported on top of it.

Pile foundations are used in the following situations:

1. When there is a layer of weak soil at the surface. This layer cannot support the weight of the building, so the loads of the building have to bypass this layer and be transferred to the layer of stronger soil or rock that is below the weak layer.
2. When a building has very heavy, concentrated loads, such as in a high rise structure.

Pile foundations are capable of taking higher loads than spread footings.

There are two types of pile foundations, each of which works in its own way.

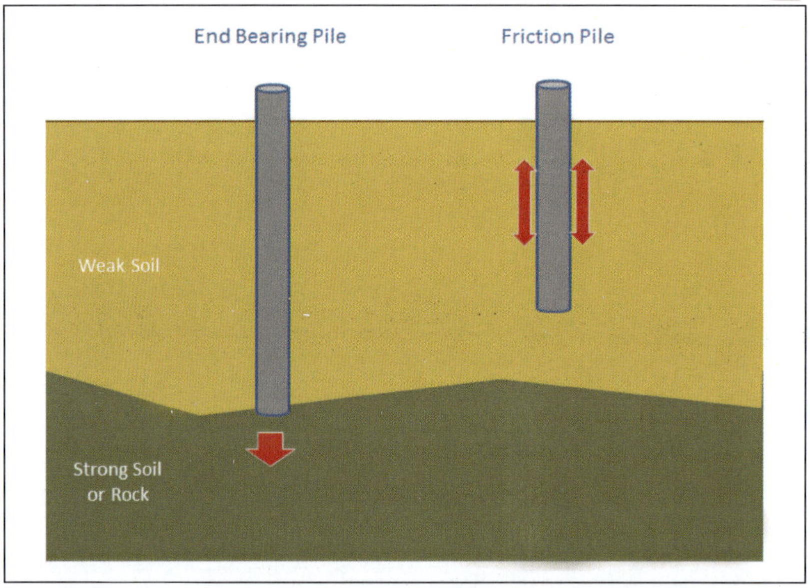

End Bearing Piles

In end bearing piles, the bottom end of the pile rests on a layer of especially strong soil or rock. The load of the building is transferred through the pile onto the strong layer. In a sense, this pile acts like a column. The key principle is that the bottom end rests on the surface which is the intersection of a weak and strong layer. The load therefore bypasses the weak layer and is safely transferred to the strong layer.

Friction Piles

Friction piles work on a different principle. The pile transfers the load of the building to the soil across the full height of the pile, by friction. In other words, the entire surface of the pile, which is cylindrical in shape, works to transfer the forces to the soil.

To visualise how this works, imagine you are pushing a solid metal rod of say 4mm diameter into a tub of frozen ice cream. Once you have pushed it in, it is strong enough to support some load. The greater the embedment depth in the ice cream, the more load it can support. This is very similar to how a friction pile works. In a friction pile, the amount of load a pile can support is directly proportionate to its length.

In practice, however, each pile resists load by a combination of end bearing and friction.

2

Construction and Building Engineering

The discipline which deals with the management, construction, planning and designing of infrastructures like airports, railroads, bridges, dams, utilities, tunnels, roads, etc. is known as construction engineering. This chapter closely examines the key concepts related to construction engineering to provide an extensive understanding of the subject.

Civil Construction

Civil construction falls in the category of civil engineering which is all about designing, constructing and maintaining the physical and naturally built environment. Civil construction is the art of building bridges, dams, roads, airports, canals, and buildings.

Civil engineering is the oldest disciplines of engineering. Since the very beginning of the human exis-tence, it has been the aspect of life. One of the main aspects of civil construction is structural engineer-ing. It helps in designing the structure in a way that it is able to support itself successfully along with resisting loads. The stability of the design will serve as the foundation of civil construction.

Civil Construction Stages

Classifying large projects into civil construction stages has always been anticipated whether it is for independent or conjunction construction. To escalate the progress of the civil construction, every stage of the process is taken as individual contracts. All of the contracts are continued in special patterns and schedules. To maintain the quality of work in the construction process, the duration of every stage should be analyzed and evaluated before the process starts. Every stage of construction should have its own time period of implementations. The stages of construction for a couple of projects are mentioned below:

Highway Project

Take a look at the factors to consider while planning a highway construction project.

- Locating the plant for the project site.
- Dissipating the site of the project.
- Underground work involving Earth work for the embankment of the highway.
- Drainage establishment.
- Pavements structure constructions.
- Placing road signs, lights and signals.

Water Supply Project

The water supply project requires construction with great care and perfection. It requires certain things that need to be taken in the spot light and they are:

- Site dissipating.
- Building of the dam structure.
- Constructing the controls and spillways.
- Pumping house connections.
- Adding in construction tunnels and transmission lines.
- Establishment of the plant of water treatment.

What does Every Construction Plan Setup Require?

To start the process of construction, the entire stages are classified further into numerous operations. Every stage of the construction process has always been facilitated with high technology equipment to sure quality and perfection. Professionals from all sorts of departments come together and work as a team to get started with these projects. Every sub operation having another operation is marked as the part of the project that is to be carried out through the division of labors with respect to skills and quality equipment. All the operations required for the process of construction of the R.C.C. retaining wall are mentioned in the list below:

- Diggings by machine
- Digging through labor force
- Assembling to form out the work
- Inserting the support
- Use of concrete and pouring it
- Filling the earth back again

Building Material

Building material is any material which is used for construction purposes. Many naturally occurring substances, such as clay, rocks, sand, and wood, even twigs and leaves, have been used to construct buildings. Apart from naturally occurring materials, many man-made products are in use, some more and some less synthetic. The manufacturing of building materials is an established industry in many countries and the use of these materials is typically segmented into specific specialty trades, such as carpentry, insulation, plumbing, and roofing work. They provide the make-up of habitats and structures including homes.

Total Cost of Building Materials

In history there are trends in building materials from being natural to becoming more man-made and composite; biodegradable to imperishable; indigenous (local) to being transported globally; repairable to disposable; chosen for increased levels of fire-safety, and improved seismic resistance. These trends tend to increase the *initial* and *long term* economic, ecological, energy, and social costs of building materials.

Economic Costs

Initial economic cost of building materials is the purchase price. This is often what governs decision making about what materials to use. Sometimes people take into consideration the energy savings or durability of the materials and see the value of paying a higher initial cost in return for a lower lifetime cost. For example, an asphalt shingle roof costs less than a metal roof to install, but the metal roof will last longer so the lifetime cost is less per year. Some materials may require more care than others, maintaining costs specific to some materials may also influence the final decision. Risks when considering lifetime cost of a material is if the building is damaged such as by fire or wind, or if the material is not as durable as advertised. The cost of materials should be taken into consideration to bear the risk to buy combustive materials to enlarge the lifetime. It is said that, 'if it must be done, it must be done well'.

Ecological Costs

Pollution costs can be macro and micro. The macro, environmental pollution of extraction industries building materials rely on such as mining, petroleum, and logging produce environmental damage at their source and in transportation of the raw materials, manufacturing, transportation of the products, retailing, and installation. An example of the micro aspect of pollution is the off-gassing of the building materials in the building or indoor air pollution. Red List building materials are materials found to be harmful. Also the carbon footprint, the total set of greenhouse gas emissions produced in the life of the material. A life-cycle analysis also includes the reuse, recycling, or disposal of construction waste. Two concepts in building which account for the ecological economics of building materials are green building and sustainable development.

Energy Costs

the Initial energy costs include the amount of energy consumed to produce, deliver and install the material. The long term energy cost is the economic, ecological, and social costs of continuing to produce and deliver energy to the building for its use, maintenance, and eventual removal. The initial embodied energy of a structure is the energy consumed to extract, manufacture, deliver, install, the materials. The lifetime embodied energy continues to grow with the use, maintenance, and reuse/recycling/disposal of the building materials themselves *and* how the materials and design help minimize the life-time energy consumption of the structure.

Social Costs

Social costs are injury and health of the people producing and transporting the materials and potential health problems of the building occupants if there are problems with the building biology. Globalization has had significant impacts on people both in terms of jobs, skills, and self-sufficiency are lost when manufacturing facilities are closed and the cultural aspects of where new facilities are opened. Aspects of fair trade and labor rights are social costs of global building material manufacturing.

Naturally Occurring Substances

Brush

Brush structures are built entirely from plant parts and were used in primitive cultures such as Native Americans and pygmy peoples in Africa These are built mostly with branches, twigs and leaves, and bark, similar to a beaver's lodge. These were variously named wikiups, lean-tos, and so forth.

An extension on the brush building idea is the wattle and daub process in which clay soils or dung, usually cow, are used to fill in and cover a woven brush structure. This gives the structure more thermal mass and strength. Wattle and daub is one of the oldest building techniques. Many older timber frame buildings incorporate wattle and daub as non load bearing walls between the timber frames.

Ice and Snow

Snow and occasionally ice, were used by the Inuit peoples for igloos and snow is used to build a shelter called a quinzhee. Ice has also been used for ice hotels as a tourist attraction in northern climates.

Mud and Clay

Clay based buildings usually come in two distinct types. One being when the walls are made directly with the mud mixture, and the other being walls built by stacking air-dried building blocks called mud bricks.

Other uses of clay in building is combined with straws to create light clay, wattle and daub, and mud plaster.

Sod buildings in Iceland.

Wet-laid Clay Walls

Wet-laid, or damp, walls are made by using the mud or clay mixture directly without forming blocks and drying them first. The amount of and type of each material in the mixture used leads to different styles of buildings. The deciding factor is usually connected with the quality of the soil being used. Larger amounts of clay are usually employed in building with cob, while low-clay soil is usually associated with sod house or sod roof construction. The other main ingredients include more or less sand/gravel and straw/grasses. Rammed earth is both an old and newer take on creating walls, once made by compacting clay soils between planks by hand; nowadays forms and mechanical pneumatic compressors are used.

Soil, and especially clay, provides good thermal mass; it is very good at keeping temperatures at a constant level. Homes built with earth tend to be naturally cool in the summer heat and warm in cold weather. Clay holds heat or cold, releasing it over a period of time like stone. Earthen walls change temperature slowly, so artificially raising or lowering the temperature can use more resources than in say a wood built house, but the heat/coolness stays longer.

People building with mostly dirt and clay, such as cob, sod, and adobe, created homes that have been built for centuries in western and northern Europe, Asia, as well as the rest of the world, and continue to be built, though on a smaller scale. Some of these buildings have remained habitable for hundreds of years.

Structural Clay Blocks and Bricks

Mud-bricks, also known by their Spanish name *adobe* are ancient building materials with evidence dating back thousands of years BC. *Compressed earth blocks* are a more modern type of brick used for building more frequently in industrialized society since the building blocks can be manufactured off site in a centralized location at a brickworks and transported to multiple building locations. These blocks can also be monetized more easily and sold.

Structural mud bricks are almost always made using clay, often clay soil and a binder are the only ingredients used, but other ingredients can include sand, lime, concrete, stone and other binders. The formed or compressed block is then air dried and can be laid dry or with a mortar or clay slip.

Sand

Sand is used with cement, and sometimes lime, to make mortar for masonry work and plaster.

Sand is also used as a part of the concrete mix. An important low-cost building material in countries with high sand content soils is the Sandcrete block, which is weaker but cheaper than fired clay bricks.

Stone or Rock

Rock structures have existed for as long as history can recall. It is the longest lasting building material available, and is usually readily available. There are many types of rock throughout the world, all with differing attributes that make them better or worse for particular uses. Rock is a very dense material so it gives a lot of protection too; its main drawback as a material is its weight and awkwardness. Its energy density is also considered a big drawback, as stone is hard to keep warm without using large amounts of heating resources.

Dry-stone walls have been built for as long as humans have put one stone on top of another. Eventually, different forms of mortar were used to hold the stones together, cement being the most commonplace now.

The granite-strewn uplands of Dartmoor National Park, United Kingdom, for example, provided ample resources for early settlers. Circular huts were constructed from loose granite rocks throughout the Neolithic and early Bronze Age, and the remains of an estimated 5,000 can still be seen today. Granite continued to be used throughout the Medieval period and into modern times. Slate is another stone type, commonly used as roofing material in the United Kingdom and other parts of the world where it is found.

Stone buildings can be seen in most major cities; some civilizations built entirely with stone such as the Egyptian and Aztec pyramids and the structures of the Inca civilization.

Thatch

Toda tribe hut.

Thatch is one of the oldest of building materials known, Thatch is another word for grass; grass is a good insulator and easily harvested. Many African tribes have lived in homes made completely of grasses and sand year-round. In Europe, thatch roofs on homes were once prevalent but the material fell out of favor as industrialization and improved transport increased the availability of other materials. Today, though, the practice is undergoing a revival. In the Netherlands, for instance, many new buildings have thatched roofs with special ridge tiles on top.

Wood and Timber

A wood-framed house under construction in Texas, United States.

Wood has been used as a building material for thousands of years in its natural state. Today, engineered wood is becoming very common in industrialized countries.

Wood is a product of trees, and sometimes other fibrous plants, used for construction purposes when cut or pressed into lumber and timber, such as boards, planks and similar materials. It is a generic building material and is used in building just about any type of structure in most climates. Wood can be very flexible under loads, keeping strength while bending, and is incredibly strong when compressed vertically. There are many differing qualities to the different types of wood, even among same tree species. This means specific species are better suited for various uses than others. And growing conditions are important for deciding quality.

The Gliwice Radio Tower (the second tallest wooden structure in the world) in Poland.

"Timber" is the term used for construction purposes except the term "lumber" is used in the United States. Raw wood (a log, trunk, bole) becomes timber when the wood has been "converted" (sawn, hewn, split) in the forms of minimally-processed logs stacked on top of each other, timber frame construction, and light-frame construction. The main problems with timber structures are fire risk and moisture-related problems.

In modern times softwood is used as a lower-value bulk material, whereas hardwood is usually used for finishings and furniture. Historically timber frame structures were built with oak in western Europe, recently douglas fir has become the most popular wood for most types of structural building.

Many families or communities, in rural areas, have a personal woodlot from which the family or community will grow and harvest trees to build with or sell. These lots are tended to like a garden. This was much more prevalent in pre-industrial times, when laws existed as to the amount of wood one could cut at any one time to ensure there would be a supply of timber for the future, but is still a viable form of agriculture.

Man-made Substances

Fired Bricks and Clay Blocks

A pile of fired bricks.

Clay blocks (sometimes called clay block brick) being laid with an adhesive rather than mortar.

Bricks are made in a similar way to mud-bricks except without the fibrous binder such as straw and are fired ("burned" in a brick clamp or kiln) after they have air-dried to permanently harden them. Kiln fired clay bricks are a ceramic material. Fired bricks can be solid or have hollow cavities to aid in drying and make them lighter and easier to transport. The individual bricks are placed upon each other in courses using mortar. Successive courses being used to build up walls, arches, and other architectural elements. Fired brick walls are usually substantially thinner than cob/ adobe while keeping the same vertical strength. They require more energy to create but are easier to transport and store, and are lighter than stone blocks. Romans extensively used fired brick of a shape and type now called Roman bricks. Building with brick gained much popularity in the mid-18th century and 19th centuries. This was due to lower costs with increases in brick manufacturing and fire-safety in the ever crowding cities.

The cinder block supplemented or replaced fired bricks in the late 20th century often being used for the inner parts of masonry walls and by themselves. Structural clay tiles (clay blocks) are clay or terracotta and typically are perforated with holes.

Cement Composites

Cement bonded composites are made of hydrated cement paste that binds wood, particles, or fibers to make pre-cast building components. Various fiberous materials, including paper, fiberglass, and carbon-fiber have been used as binders.

Wood and natural fibers are composed of various soluble organic compounds like carbohydrates, glycosides and phenolics. These compounds are known to retard cement setting. Therefore, before using a wood in making cement bonded composites, its compatibility with cement is assessed.

Wood-cement compatibility is the ratio of a parameter related to the property of a wood-cement composite to that of a neat cement paste. The compatibility is often expressed as a percentage value. To determine wood-cement compatibility, methods based on different properties are used, such as, hydration characteristics, strength, interfacial bond and morphology. Various methods are used by researchers such as the measurement of hydration characteristics of a cement-aggregate mix; the comparison of the mechanical properties of cement-aggregate mixes and the visual assessment of microstructural properties of the wood-cement mixes. It has been found that the hydration test by measuring the change in hydration temperature with time is the most convenient method. Recently, Karade et al. have reviewed these methods of compatibility assessment and suggested a method based on the 'maturity concept' i.e. taking in consideration both time and temperature of cement hydration reaction.

Bricks were laid in lime mortar from the time of the Romans until supplanted by Portland cement mortar in the early 20th century. Cement blocks also sometimes are filled with grout or covered with a parge coat.

Concrete

Falkirk Wheel.

Concrete is a composite building material made from the combination of aggregate and a binder such as cement. The most common form of concrete is Portland cement concrete, which consists of mineral aggregate (generally gravel and sand), portland cement and water.

After mixing, the cement hydrates and eventually hardens into a stone-like material. When used in the generic sense, this is the material referred to by the term "concrete".

For a concrete construction of any size, as concrete has a rather low tensile strength, it is generally strengthened using steel rods or bars (known as rebars). This strengthened concrete is then referred to as reinforced concrete. In order to minimise any air bubbles, that would weaken the structure, a vibrator is used to eliminate any air that has been entrained when the liquid concrete mix is poured around the ironwork. Concrete has been the predominant building material in the modern age due to its longevity, formability, and ease of transport. Recent advancements, such as insulating concrete forms, combine the concrete forming and other construction steps (installation of insulation). All materials must be taken in required proportions as described in standards.

Fabric

The tent is the home of choice among nomadic groups all over the world. Two well-known types include the conical teepee and the circular yurt. The tent has been revived as a major construction technique with the development of tensile architecture and synthetic fabrics. Modern buildings can be made of flexible material such as fabric membranes, and supported by a system of steel cables, rigid or internal, or by air pressure.

Foam

Foamed plastic sheet to be used as backing for firestop mortar at CIBC bank in Toronto.

Recently, synthetic polystyrene or polyurethane foam has been used in combination with structural materials, such as concrete. It is lightweight, easily shaped, and an excellent insulator. Foam is usually used as part of a structural insulated panel, wherein the foam is sandwiched between wood or cement or insulating concrete forms.

Glass

Glassmaking is considered an art form as well as an industrial process or material.

Clear windows have been used since the invention of glass to cover small openings in a building. Glass panes provided humans with the ability to both let light into rooms while at the same time keeping inclement weather outside.

Glass is generally made from mixtures of sand and silicates, in a very hot fire stove called a kiln, and is very brittle. Additives are often included the mixture used to produce glass with shades of colors or various characteristics such as bulletproof glass or lightbulbs.

The use of glass in architectural buildings has become very popular in the modern culture. Glass "curtain walls" can be used to cover the entire facade of a building, or it can be used to span over a wide roof structure in a "space frame". These uses though require some sort of frame to hold sections of glass together, as glass by itself is too brittle and would require an overly large kiln to be used to span such large areas by itself.

Gypcrete

Gypcrete is a mixture of gypsum plaster and fibreglass rovings. Although plaster and fibres fiborous plaster have been used for many years, especially for ceilings, it was not until the early 1990s that serious studies of the strength and qualities of a walling system Rapidwall, using a mixture of gypsum plaster and 300mm plus fibreglass rovings, were investigated. It was discovered, through testing at the University of Adelaide, that these walls had significant, load bearing, shear and lateral resistance together with earthquake-resistance, fire-resistance, and thermal properties. With an abundance of gypsum (naturally occurring and by-product chemical FGD and phospho gypsums) available worldwide, gypcrete-based building products, which are fully recyclable, offer significant environmental benefits.

Metal

Metal is used as structural framework for larger buildings such as skyscrapers, or as an external surface covering. There are many types of metals used for building. Metal figures quite prominently in prefabricated structures such as the Quonset hut, and can be seen used in most cosmopolitan cities. It requires a great deal of human labor to produce metal, especially in the large amounts needed for the building industries. Corrosion is metal's prime enemy when it comes to longevity.

- Steel is a metal alloy whose major component is iron, and is the usual choice for metal structural building materials. It is strong, flexible, and if refined well and treated lasts a long time.
- The lower density and better corrosion resistance of aluminium alloys and tin sometimes overcome their greater cost.

Copper belfry of St. Laurentius church, Bad Neuenahr-Ahrweiler.

- Copper is a valued building material because of its advantageous properties. These include corrosion resistance, durability, low thermal movement, light weight, radio frequency shielding, lightning protection, sustainability, recyclability, and a wide range of finishes. Copper is incorporated into roofing, flashing, gutters, downspouts, domes, spires, vaults, wall cladding, building expansion joints, and indoor design elements.

- Other metals used include chrome, gold, silver, and titanium. Titanium can be used for structural purposes, but it is much more expensive than steel. Chrome, gold, and silver are used as decoration, because these materials are expensive and lack structural qualities such as tensile strength or hardness.

Plastics

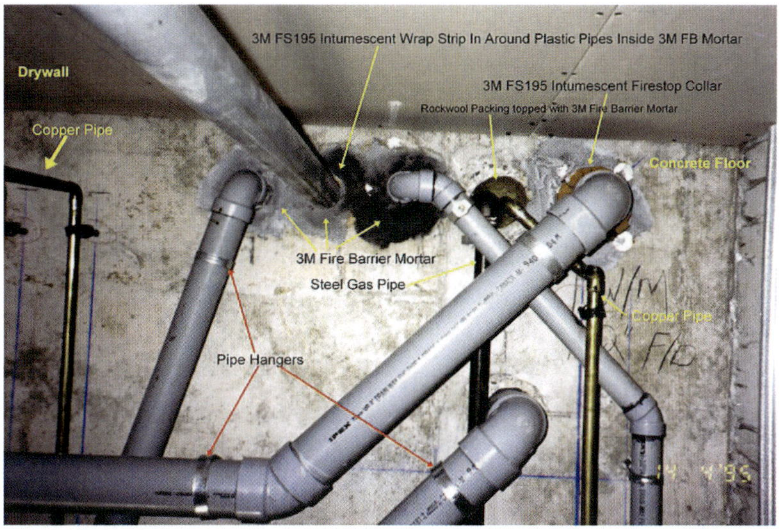

Plastic pipes penetrating a concrete floor in a Canadian highrise apartment building.

The term "plastics" covers a range of synthetic or semi-synthetic organic condensation or polymerization products that can be molded or extruded into objects, films, or fibers. Their name is derived from the fact that in their semi-liquid state they are malleable, or have the property of plasticity. Plastics vary immensely in heat tolerance, hardness, and resiliency. Combined with this adaptability, the general uniformity of composition and lightness of plastics ensures their use in almost all industrial applications today. High performance plastics such as ETFE have become an ideal building material due to its high abrasion resistance and chemical inertness. Notable buildings that feature it include: the Beijing National Aquatics Center and the Eden Project biomes.

Papers and Membranes

Building papers and membranes are used for many reasons in construction. One of the oldest building papers is red rosin paper which was known to be in use before 1850 and was used as an underlayment in exterior walls, roofs, and floors and for protecting a jobsite during construction. Tar paper was invented late in the 19th century and was used for similar purposes as rosin paper and for gravel roofs. Tar paper has largely fallen out of use supplanted by asphalt felt paper. Felt paper has been supplanted in some uses by synthetic underlayments, particularly in roofing by synthetic underlayments and siding by housewraps. There are a wide variety of damp proofing and waterproofing membranes used for roofing, basement waterproofing, and geomembranes.

Ceramics

Fired clay bricks have been used since the time of the Romans. Special tiles are used for roofing, siding, flooring, ceilings, pipes, flue liners, and more.

Building Products

In the market place the term “building products” often refers to ready-made particles/sections, made from various materials, that are fitted in architectural hardware and decorative hardware parts of a building. The list of building products excludes the building materials used to construct the building architecture and supporting fixtures, like windows, doors, cabinets, etc. Building products, rather, support and make building materials work in a modular fashion.

“Building products” may also refer to items used to put such hardware together, such as caulking, glues, paint, and anything else bought for the purpose of constructing a building.

Sustainability

In 2017, the buildings and construction together consumed 36% of the final energy produced globally while being responsible for 39% of the global energy related CO_2 emissions. The shares from construction industry alone were 6% and 11% respectively. Energy consumption during the building-material production, predominantly due to their use of electricity, is a dominant contributor to the construction industry’s share. Embodied energy of relevant building materials in the US are provided in the table below.

Material	Embodied energy	
	kBtu/lb	MJ/kg
bricks	1.66	3.86116
cement	3.23	7.51298
clay	15.2	35.3552
concrete	0.58	1.34908
copper	25.77	59.94102
flat glass	10.62	24.70212
gypsum	10.38	24.14388
hardwood plywood & veneer	15.19	35.33194
lime	1.92	4.46592
mineral wool insulation	12.6	29.3076
primary aluminum	80.17	186.47542
softwood plywood & veneer	3.97	9.23422
stone	1.43	3.32618
virgin steel	10.39	24.16714
wood lumber	2.7	6.2802

Building Stones

Stones are derived from the rocks which are naturally occurred as the portion of earth’s crust. Each rock has a different chemical composition and exhibits different properties. Some of these rocks are widely used in the construction industry.

Commonly used Building Stones

Following are the commonly used building stones:

- Granite
- Basalt and trap
- Serpentine
- Limestone
- Chalk
- Sandstone
- Caliche
- Marble
- Slate
- Quartzite
- Laterite
- Gneiss

Granite

Granite is an igneous rock which is formed by the solidification of magma at a considerable depth from the earth's surface. It is hard and durable. It is available in different colors. The color of granite varies according to the amount of feldspar in it. It can be polished nicely.

The compressive strength of granite is about 75 to 130 MPa. Its specific gravity lies between 2.63 to 2.75. Its density is about 26 to 27 KN/m^3.

Uses

Granite is used for large engineering projects such as the construction of bridge abutments, dams, offshore structures, etc. It is also used for constructing steps, walls, flooring, etc.

Granite Tiles.

Basalt and Trap

Basalt and trap are also igneous rocks. These stones are also called as green stones or white stones or blue basalt. They are hard and tough. It is difficult to work with this kind of rocks. The specific gravity of basalt and trap varies from 2.6 to 3.0. They are having a high compressive strength of 150 to 190 MPa.

Uses

Basalt and trap used as railway ballast, aggregate in concrete, pavement material, etc. Yellow and red-colored basalt and trap are used to construct decorative features in structures.

Basalt and Trap as Railway Ballast.

Syenite

Syenite is also an igneous rock which is hard and durable like granite. The structure of syenite is coarse-grained and its crushing strength varies from 90 to 150 MPa. It is available in different colors but typically they are light in color.

Uses of Syenite

Crushed syenite is commonly used as aggregate in pavement construction and concrete making. It is also used for external facings of building walls etc.

Syenite.

Limestone

Limestone belongs to sedimentary rocks which are formed by the deposition of particles of weathering. It contains calcium carbonate in huge proportion. It is soft and easily workable. The compressive strength of limestone is about 55 MPa and specific gravity lies between 2.0 to 2.75.

Uses

Limestone is used to manufacture lime and cement. It is also used for flooring, roofing, etc.

Limestone Blocks.

Chalk

Chalk is a sedimentary rock and it contains pure lime in it. It is very soft and easily can be converted into powdered form. It is generally white in color and contains a porous structure.

Uses of Chalk

powdered chalk is used for manufacturing lime putty and also used as a pigment in Portland cement.

Chalk Rocks.

Sandstone

Sandstone is a sedimentary rock and consists of different minerals like quartz, feldspar, silica, etc. in it. The hardness of sandstone depends upon the proportion of minerals in it. However, it is easily workable and also available in different colors. The specific gravity of sandstone lies between 2.65 and 2.95. Its compressive strength is about 65 MPa and with a density of 20 to 22 KN/m_3.

Uses of Sandstone

It is used as material for roofing, paving, columns, facing works and ornamental carvings, etc.

Sandstone Columns.

Caliche

Caliche also called as Kankar falls under the sedimentary category of rocks. It is the impure form of limestone. It is porous in structure and irregular in shape. Its hardness varies from soft to hard.

Uses of Caliche

Caliche or kankar is used as aggregate in pavement construction. It is also used to prepare hydraulic lime.

Kankar Bedding Planes.

Marble

Marble belongs to the metamorphic category of rocks and it is formed when limestone is subjected to excessive heat and pressure. This process of forming is called metamorphism. Marble is hard and compact in nature. It occurs in different colors and also it can take a good polish. Its compressive strength is about 70 Mpa. The specific gravity of marble is about 2.65.

Uses of Marble

Marble Flooring.

Marble is used for flooring, facing works, steps, etc. It can be carved into required shape easily hence, it is used for decorative and ornamental works of structures.

Slate

Slate is a metamorphic rock which is formed by the metamorphic action of shale under low pressure and heat conditions. It is hard and brittle. It consists of foliated texture and generally black in color. It can be split into thin slabs along its natural foliated planes. Its crushing strength ranges from 75 to 210 Mpa with a specific gravity of 2.89.

Uses of Slate

Slate is used for floorings, roofing works, partitions, damp proof courses, etc.

Slate Roofing.

Quartzite

Quartzite also belongs metamorphic category which is formed by the metamorphism of sandstone under high pressure and temperature. It is of crystalline structure with a granular texture. It is hard and brittle. It is not easily workable. Its compressive strength is about 115 MPa.

Uses of Quartzite

Quartzite is used as road aggregate, concrete aggregate, etc. it is also used to construct retaining walls, rubble masonry, stone pitching, etc.

Quartzite Retaining Wall.

Laterite

Laterite also belongs to the metamorphic category. It is porous and spongy in its structure. It is soft and workable hence quarrying of laterite is easy. It has low crushing strength which is about 1.8 to 3.0 MPa. When it is decomposed it turns into moorum which is widely used for road construction.

Uses Laterite

Laterite is used for rough stone masonry works, pavement construction works, etc.

Laterite Masonry Blocks.

Gneiss

Gneiss is another metamorphic rock formed by the metamorphosis of granite. It consists of foliated structure hence it can be split into thin slabs along its bedding planes. It is strong and durable. Its crushing strength varies from 206 MPa to 370 MPa with a specific gravity of 2.69.

Uses of Gneiss

Crushed gneiss used for pavement construction, rough stone masonry works, stone pitching, etc.

Gneiss Stones.

Construction Engineering

Construction engineering is a professional discipline that deals with the designing, planning, construction and management of infrastructures such as roads, tunnels, bridges, airports, railroads, facilities, buildings, dams, utilities and other projects.

Civil engineering is a related field that deals more with the practical aspects of projects. Construction engineers learn some of the design aspects similar to civil engineers as well as project site management aspects.

At the educational level, civil engineering students concentrate primarily on the design work which is more analytical, gearing them toward a career as a design professional. This essentially requires them to take a multitude of challenging engineering science and design courses as part of obtaining a 4-year accredited degree. Education for construction engineers is primarily focused on construction procedures, methods, costs, schedules and personnel management. Their primary concern is to deliver a project on time within budget and of the desired quality.

The difference between a construction engineer and a civil engineer is that construction engineering students take basic design courses as well as construction management courses.

Abilities

Construction engineers are problem solvers. They contribute to the creation of infrastructure that best meets the unique demands of its environment. They must be able to understand infrastructure life cycles. When compared and contrasted to design engineers, construction engineers bring to the table their own unique perspectives for solving technical challenges with clarity and imagination. While individuals considering this career path should certainly have a strong understanding of mathematics and science, many other skills are also highly desirable, including critical and analytical thinking, time management, people management and good communication skills.

Architecture Engineering

Architectural engineering, also known as building engineering or architecture engineering, is a discipline that deals with the technological aspects and multi-disciplinary approach to planning, design, construction and operation of buildings, such as analysis and integrated design of environmental systems (energy conservation, HVAC, lighting, acoustics, vertical and horizontal transportation), structural systems, behavior and properties of building components and materials, and construction management.

From reduction of greenhouse gas emissions to the construction of resilient buildings, architectural engineers are at the forefront of addressing several major challenges of the 21st century. They apply the latest scientific knowledge and technologies to the design of buildings. Architectural engineering as a relatively new licensed profession emerged in the 20th century as a result of the rapid technological developments. Architectural engineers are at the forefront of two major historical opportunities that today's world is immersed in: (1) that of rapidly advancing computer-technology, and (2) the parallel revolution arising from the need to create a sustainable planet.

Distinguished from architecture as an art of design, architectural Engineering, is the art and science of engineering and construction as practiced in respect of buildings.

Related Engineering and Design Fields

Structural Engineering

Structural engineering involves the analysis and design of the built environment (buildings, bridges, equipment supports, towers and walls). Those concentrating on buildings are sometimes informally referred to as "building engineers". Structural engineers require expertise in strength of materials, structural analysis, and in predicting structural load such as from weight of the building, occupants and contents, and extreme events such as wind, rain, ice, and seismic design of structures which is referred to as earthquake engineering. Architectural Engineers sometimes incorporate structural as one aspect of their designs; the structural discipline when practiced as a specialty works closely with architects and other engineering specialists.

Mechanical, Electrical and Plumbing (MEP)

MEP room in a building.

Mechanical engineering and electrical engineering engineers are specialists, commonly referred to as (Mechanical, Electrical and Plumbing) when engaged in the building design fields. Also known as "building services engineering" in the United Kingdom, Canada and Australia. Mechanical engineers often design and oversee the heating, ventilation and air conditioning (HVAC), plumbing and rain gutter systems. Plumbing designers often include design specifications for simple active fire protection systems, but for more complicated projects, fire protection engineers are often separately retained. Electrical engineers are responsible for the building's power distribution, telecommunication, fire alarm, signalization, lightning protection and control systems, as well as lighting systems.

Construction Management

Construction management (CM) is a professional service that uses specialized, project management techniques to oversee the planning, design, and construction of a project, from its beginning to its end. The purpose of CM is to control a project's time/delivery, cost and quality—sometimes referred to as a project management triangle or "triple constraints." CM is compatible with all project delivery systems, including design-bid-build, design-build, CM At-Risk and Public Private Partnerships. Professional construction managers may be reserved for lengthy, large-scale, high budget undertakings (commercial real estate, transportation infrastructure, industrial facilities, and military infrastructure), called capital projects.

The Role of a Contractor

Contractors are assigned to a construction project during the design or once the design has been completed by a licensed architect or a licensed civil engineer. This is done by going through a bidding process with different contractors. The contractor is selected by using one of three common selection methods: low-bid selection, best-value selection, or qualifications-based selection.

A construction manager should have the ability to handle public safety, time management, cost management, quality management, decision making, mathematics, working drawings, and human resources.

Function

The functions of construction management typically include the following:

- Specifying project objectives and plans including delineation of scope, budgeting, scheduling, setting performance requirements, and selecting project participants.
- Maximizing the resource efficiency through procurement of labor, materials and equipment.
- Implementing various operations through proper coordination and control of planning, design, estimating, contracting and construction in the entire process.
- Developing effective communications and mechanisms for resolving conflicts.

The Construction Management Association of America (CMAA) states the most common responsibilities of a Construction Manager fall into the following 7 categories: Project Management Planning, Cost Management, Time Management, Quality Management, Contract Administration, Safety Management, and CM Professional Practice. CM professional practice includes specific activities, such as defining the responsibilities and management structure of the project management team, organizing and leading by implementing project controls, defining roles and responsibilities, developing communication protocols, and identifying elements of project design and construction likely to give rise to disputes and claims.

Seven Types of Construction

- Agricultural: Typically economical buildings, and other improvements, for agricultural purposes. Examples include barns, equipment and animal sheds, specialized fencing, storage silos and elevators, and water supply and drains such as wells, tanks, and ditches.
- Residential: Residential construction includes houses, apartments, townhouses, and other smaller, low-rise housing, small office types.
- Commercial: This refers to construction for the needs of private commerce, trade, and services. Examples include office buildings, "big box" stores, shopping centers and malls, warehouses, banks, theaters, casinos, resorts, golf courses, and larger residential structures such as high-rise hotels and condominiums.

- Institutional: This category is for the needs of government and other public organizations. Examples include schools, fire and police stations, libraries, museums, dormitories, research buildings, hospitals, transportation terminals, some military facilities, and governmental buildings.
- Industrial: Buildings and other constructed items used for storage and product production, including chemical and power plants, steel mills, oil refineries and platforms, manufacturing plants, pipelines, and seaports.
- Heavy civil: The construction of transportation infrastructure such as roads, bridges, railroads, tunnels, airports, and fortified military facilities. Dams are also included, but most other water-related infrastructure is considered environmental.
- Environmental: Environmental construction was part of heavy civil, but is now separate, dealing with projects that improve the environment. Some examples are water and wastewater treatment plants, sanitary and storm sewers, solid waste management, and air pollution control.

Obtaining the Project

Bids

A bid is given to the owner by construction managers that are willing to complete their construction project. A bid tells the owner how much money they should expect to pay the construction management company in order for them to complete the project.

- Open bid: An open bid is used for public projects. Any and all contractors are allowed to submit their bid due to public advertising.
- Closed bid: A closed bid is used for private projects. A selection of contractors are sent an invitation for bid so only they can submit a bid for the specified project.

Selection Methods

- Low-bid selection: This selection focuses on the price of a project. Multiple construction management companies submit a bid to the owner that is the lowest amount they are willing to do the job for. Then the owner usually chooses the company with the lowest bid to complete the job for them.
- Best-value selection: This selection focuses on both the price and qualifications of the contractors submitting bids. This means that the owner chooses the contractor with the best price and the best qualifications. The owner decides by using a request for proposal (RFP), which provides the owner with the contractor's exact form of scheduling and budgeting that the contractor expects to use for the project.
- Qualifications-based selection: This selection is used when the owner decides to choose the contractor only on the basis of their qualifications. The owner then uses a request for qualifications (RFQ), which provides the owner with the contractor's experience, management plans, project organization, and budget and schedule performance. The owner may also

ask for safety records and individual credentials of their members. This method is most often used when the contractor is hired early during the design process so that the contractor can provide input and cost estimates as the design develops.

Payment Contracts

- Lump sum: This is the most common type of contract. The construction manager and the owner agree on the overall cost of the construction project and the owner is responsible for paying that amount whether the construction project exceeds or falls below the agreed price of payment.
- Cost plus fee: This contract provides payment for the contractor including the total cost of the project as well as a fixed fee or percentage of the total cost. This contract is beneficial to the contractor since any additional costs will be paid for, even though they were unexpected for the owner.
- Guaranteed maximum price: This contract is the same as the cost-plus-fee contract although there is a set price that the overall cost and fee do not go above.
- Unitprice: This contract is used when the cost cannot be determined ahead of time. The owner provides materials with a specific unit price to limit spending.

Project Stages

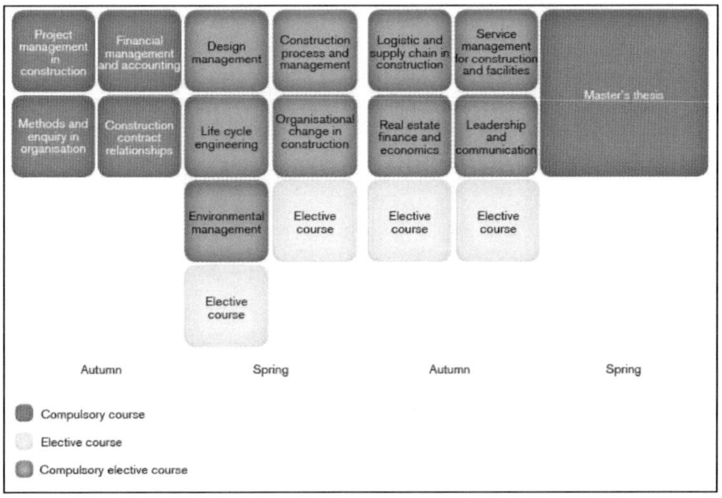

Characteristics of construction project stages of various project management approaches.

The stages of a typical construction project have been defined as feasibility, design, construction and operation, each stage relating to the project life cycle.

Feasibility and Design

Feasibility and design involves four steps: programming and feasibility, schematic design, design development, and contract documents. It is the responsibility of the design team to ensure that the design meets all building codes and regulations. It is during the design stage that the bidding process takes place.

- Conceptual/Programming and feasibility: The needs, goals, and objectives must be determined for the building. Decisions must be made on the building size, number of rooms, how the space will be used, and who will be using the space. This must all be considered to begin the actual designing of the building. This phase is normally a written list of each room or space, the critical information about those spaces, and the approximate square footage of each area.
- Schematic design: Schematic designs are sketches used to identify spaces, shapes, and patterns. Materials, sizes, colors, and textures must be considered in the sketches. This phase usually involves developing the floor plan, elevations, a site plan, and possibly a few details.
- Design development (DD): This step requires research and investigation into what materials and equipment will be used as well as their cost. During this phase, the drawings are refined with information from structural, plumbing, mechanical, and electrical engineers. It also involves a more rigorous evaluation how the applicable building codes will impact the project.
- Contract documents (CDs): Contract documents are the final drawings and specifications of the construction project. They are used by contractors to determine their bid while builders use them for the construction process. Contract documents can also be called working drawings.

Pre-construction

The pre-construction stage begins when the owner gives a notice to proceed to the contractor that they have chosen through the bidding process. A notice to proceed is when the owner gives permission to the contractor to begin their work on the project. The first step is to assign the project team which includes the project manager (PM), contract administrator, superintendent, and field engineer.

- Project manager: The project manager is in charge of the project team.
- Contract administrator: The contract administrator assists the project manager as well as the superintendent with the details of the construction contract.
- Superintendent: It is the superintendent's job to make sure everything is on schedule including flow of materials, deliveries, and equipment. They are also in charge of coordinating on-site construction activities.
- Field engineer: A field engineer is considered an entry-level position and is responsible for paperwork.

During the pre-construction stage, a site investigation must take place. A site investigation takes place to discover if any steps need to be implemented on the job site. This is in order to get the site ready before the actual construction begins. This also includes any unforeseen conditions such as historical artifacts or environment problems. A soil test must be done to determine if the soil is in good condition to be built upon.

Procurement

The procurement stage is when labor, materials and equipment needed to complete the project are purchased. This can be done by the general contractor if the company does all their own construction work. If the contractor does not do their own work, they obtain it through subcontractors. Subcontractors are contractors who specialize in one particular aspect of the construction work such as concrete, welding, glass, or carpentry. Subcontractors are hired the same way a general contractor would be, which is through the bidding process. Purchase orders are also part of the procurement stage.

- Purchase orders: A purchase order is used in various types of businesses. In this case, a purchase order is an agreement between a buyer and seller that the products purchased meet the required specifications for the agreed price.

Construction

The construction stage begins with a pre-construction meeting brought together by the superintendent (on an American project). The pre-construction meeting is meant to make decisions dealing with work hours, material storage, quality control, and site access. The next step is to move everything onto the construction site and set it all up.

A Contractor progress payment schedule is a schedule of when (according to project milestones or specified dates) contractors and suppliers will be paid for the current progress of installed work.

Progress payments are partial payments for work completed during a portion, usually a month, during a construction period. Progress payments are made to general contractors, subcontractors, and suppliers as construction projects progress. Payments are typically made on a monthly basis but could be modified to meet certain milestones. Progress payments are an important part of contract administration for the contractor. Proper preparation of the information necessary for payment processing can help the contractor financially complete the project.

Owner Occupancy

Once the owner moves into the building, a warranty period begins. This is to ensure that all materials, equipment, and quality meet the expectations of the owner that are included within the contract.

Issues Resulting from Construction

Dust and Mud

When construction vehicles are driving around a site or moving earth, a lot of dust is created, especially during the dryer months. This may cause disruption for surrounding businesses or homes. A popular method of dust control is to have a water truck driving through the site spraying water on the dry dirt to minimize the movement of dust within and out of the construction site. When water is introduced mud is created. This mud sticks to the tires of the construction vehicles and is often lead out to the surrounding roads. A street sweeper may clean the roads to reduce dirty road conditions.

Environmental Protections

- Storm water pollution: As a result of construction, the soil is displaced from its original location which can possibly cause environmental problems in the future. Runoff can occur during storms which can possibly transfer harmful pollutants through the soil to rivers, lakes, wetlands, and coastal waters.

- Endangered species: If endangered species have been found on the construction site, the site must be shut down for some time. The construction site must be shut down for as long as it takes for authorities to make a decision on the situation. Once the situation has been assessed, the contractor makes the appropriate accommodations to not disturb the species.
- Vegetation: There may often be particular trees or other vegetation that must be protected on the job site. This may require fences or security tape to warn builders that they must not be harmed.
- Wetlands: The contractor must make accommodations so that erosion and water flow are not affected by construction. Any liquid spills must be maintained due to contaminants that may enter the wetland.
- Historical or cultural artifacts: Artifacts may include arrowheads, pottery shards, and bones. All work comes to a halt if any artifacts are found and will not resume until they can be properly examined and removed from the area.

Construction Activity Documentation

Project meetings take place at scheduled intervals to discuss the progress on the construction site and any concerns or issues. The discussion and any decisions made at the meeting must be documented.

Diaries, logs, and daily field reports keep track of the daily activities on a job site each day.

- Diaries: Each member of the project team is expected to keep a project diary. The diary contains summaries of the day's events in the member's own words. They are used to keep track of any daily work activity, conversations, observations, or any other relevant information regarding the construction activities. Diaries can be referred to when disputes arise and a diary happens to contain information connected with the disagreement. Diaries that are handwritten can be used as evidence in court.
- Logs: Logs keep track of the regular activities on the job site such as phone logs, transmittal logs, delivery logs, and RFI (Request for Information) logs.
- Daily field reports: Daily field reports are a more formal way of recording information on the job site. They contain information that includes the day's activities, temperature and weather conditions, delivered equipment or materials, visitors on the site, and equipment used that day.

Labor statements are required on a daily basis. Also list of Labor, PERT CPM are needed for labor planning to complete a project in time.

Resolving Disputes

- Mediation: Mediation uses a third party mediator to resolve any disputes. The mediator helps both disputing parties to come to a mutual agreement. This process ensures that no attorneys become involved in the dispute and is less time-consuming.

- Minitrial: A minitrial takes more time and money than a mediation. The minitrial takes place in an informal setting and involves some type of advisor or attorney that must be paid. The disputing parties may come to an agreement or the third party advisor may offer their advice. The agreement is nonbinding and can be broken.
- Arbitration: Arbitration is the most costly and time-consuming way to resolve a dispute. Each party is represented by an attorney while witnesses and evidence are presented. Once all information is provided on the issue, the arbitrator makes a ruling which provides the final decision. The arbitrator provides the final decision on what must be done and it is a binding agreement between each of the disputing parties.

Software

Capital project management software (CPMS) refers to the systems that are currently available that help capital project owner/operators, program managers, and construction managers, control and manage the vast amount of information that capital construction projects create. A collection, or portfolio of projects only makes this a bigger challenge. These systems go by different names: capital project management software, computer construction software, construction management software, project management information systems. Usually Construction Management can be referred as subset of CPMS where the scope of CPMS is not limited to construction phases of project. Among main construction management software can be mentioned Procore and PlanGrid.

Required Knowledge

- Construction and building
- Technology
- Public safety
- Customer service
- Human resources
- Mathematics
- Leadership

Business Model

The construction industry typically includes three parties: an owner, a licensed designer (architect or engineer) and a builder (usually known as a general contractor). There are traditionally two contracts between these parties as they work together to plan, design and construct the project. The first contract is the owner-designer contract, which involves planning, design, and construction contract administration. The second contract is the owner-contractor contract, which involves construction. An indirect third-party relationship exists between the designer and the contractor, due to these two contracts.

Skyscrapers under construction in Panama City, Panama.

An owner may also contract with a construction project management company as an adviser, creating a third contract relationship in the project. The construction manager's role is to provide construction advice to the designer, design advice to the constructor on the owner's behalf and other advice as necessary.

Design, Bid and Build Contracts

The phrase "design, bid, build" describes the prevailing model of construction management, in which the general contractor is engaged through a tender process after designs have been completed by the architect or engineer.

Design-build Contracts

Many owners – particularly government agencies – let out contracts known as *design-build contracts*. In this type of contract, the construction team (known as the design-builder) is responsible for taking the owner's concept and completing a detailed design before (following the owner's approval of the design) proceeding with construction. Virtual design and construction technology may be used by contractors to maintain a tight construction time.

There are three main advantages to a design-build contract. First, the construction team is motivated to work with the architect to develop a practical design. The team can find creative ways to reduce construction costs without reducing the function of the final product. The second major advantage involves the schedule. Many projects are commissioned within a tight time frame. Under a traditional contract, construction cannot begin until after the design is finished and the project has been awarded to a bidder. In a design-build contract the contractor is established at the outset, and construction activities can proceed concurrently with the design. The third major advantage is that the design-build contractor has an incentive to keep the combined design and construction costs within the owner's budget. If speed is important, design and construction contracts can be awarded separately; bidding takes place on preliminary plans in a not-to-exceed contract instead of a single firm design-build contract.

The major problem with design-build contracts is an inherent conflict of interest. In a standard contract the architect works for the owner and is directly responsible to the owner. In design-build the architect works for the design-builder, not the owner, therefor the design-builder may make design and construction decisions that benefit the design-builder, but that do not benefit the owner. During construction, the architect normally acts as the owner's representative. This includes reviewing the builder's work and ensuring that the products and methods meet specifications and codes. The architect's role is compromised when the architect works for the design-builder and not for the owner directly. Thus, the owner may get a building that is over-designed to increase profits for the design-builder, or a building built with lesser-quality products to maximize profits.

Planning and Scheduling

Project-management methodology is as follows:

- Work breakdown structure
- Project network of activities
    - Critical path method (CPM)
    - Resource management
    - Resource leveling
    - Risk assessment

Architecture Engineer

- Work inspection
- Change orders
- Review payments
- Materials and samples
- Shop drawings
- Three-dimensional image

Agency CM

Construction cost management is a fee-based service in which the construction manager (CM) is responsible exclusively to the owner, acting in the owner's interests at every stage of the project. The construction manager offers impartial advice on matters such as:

- Optimum use of available funds.
- Control of the scope of the work.
- Project scheduling.
- Optimum use of design and construction firms' skills and talents.

- Avoidance of delays, changes and disputes.
- Enhancing project design and construction quality.
- Optimum flexibility in contracting and procurement.
- Cash-flow management.

Comprehensive management of every stage of the project, beginning with the original concept and project definition, yields the greatest benefit to owners. As time progresses beyond the pre-design phase, the CM's ability to effect cost savings diminishes. The agency CM can represent the owner by helping select the design and construction teams and managing the design (preventing scope creep), helping the owner stay within a predetermined budget with value engineering, cost-benefit analysis and best-value comparisons. The software-application field of construction collaboration technology has been developed to apply information technology to construction management.

CM at-risk

CM at-risk is a delivery method which entails a commitment by the construction manager to deliver the project within a Guaranteed Maximum Price (GMP). The construction manager acts as a consultant to the owner in the development and design phases (preconstruction services), and as a general contractor during construction. When a construction manager is bound to a GMP, the fundamental character of the relationship is changed. In addition to acting in the owner's interest, the construction manager must control construction costs to stay within the GMP.

CM at-risk is a global term referring to the business relationship of a construction contractor, owner and architect (or designer). Typically, a CM at-risk arrangement eliminates a "low-bid" construction project. A GMP agreement is a typical part of the CM-and-owner agreement (comparable to a "low-bid" contract), but with adjustments in responsibility for the CM. The advantage of a CM at-risk arrangement is budget management. Before a project's design is completed (six to eighteen months of coordination between designer and owner), the CM is involved with estimating the cost of constructing a project based on the goals of the designer and owner (design concept) and the project's scope. In balancing the costs, schedule, quality and scope of the project, the design may be modified instead of redesigned; if the owner decides to expand the project, adjustments can be made before pricing. To manage the budget before design is complete and construction crews mobilized, the CM conducts site management and purchases major items to efficiently manage time and cost.

Advantages

- CM is working "at risk", therefore have incentive to act in the owner's interest, as well as to efficiently manage construction costs, considering they would be liable for any amount in excess of the GMP.
- Ability to handle changes in design or scope.

Drawbacks

- If a cost overrun occurred, it could cost the CM a great deal of money.

- The CM is allowed some mistake-related contingency, therefore there is a possibility that they will compensate by reducing the scope of the work to fit the GMP.
- Since the GMP is settled before design begins, it is difficult for owners to know whether they received the best possible bid.

Bottom Line

An at-risk delivery method is best for large projects—both complete construction and renovation—that are not easy to define, have a possibility of changing in scope, or have strict schedule deadlines. Additionally, it is an efficient method in projects containing technical complexity, multi-trade coordination, or multiple phases.

Accelerated Construction Techniques

Starting with its Accelerated Bridge Program in the late 2000s, the Massachusetts Department of Transportation began employing accelerated construction techniques, in which it signs contracts with incentives for early completion and penalties for late completion, and uses intense construction during longer periods of complete closure to shorten the overall project duration and reduce cost.

Green Building Construction

A green building has four main elements or components on which it is designed: materials, energy, water and health to make green building more sustainable.

Elements of Green Building Design

Following are the components of a Green Building to make it sustainable:

Materials for Green Building

Materials for a green building are obtained from natural, renewable sources that have been managed and harvested in a sustainable way; or they are obtained locally to reduce the embedded energy costs of transportation; or salvaged from reclaimed materials at nearby sites.

Parasoleil Recycled Content Panelsclip.

Materials are assessed using green specifications that look at their Life Cycle Analysis (LCA) in terms of their embodied energy, durability, recycled content, waste minimisation, and their ability to be reused or recycled.

Chocolate bamboo.

Energy Systems in Green Buildings

Passive solar design will dramatically reduce the heating and cooling costs of a building, as will high levels of insulation and energy-efficient windows. Natural daylight design reduces a building's electricity needs, and improves people's health and productivity.

Green buildings also incorporate energy-efficient lighting, low energy appliances, and renewable energy technologies such as wind turbines and solar panels.

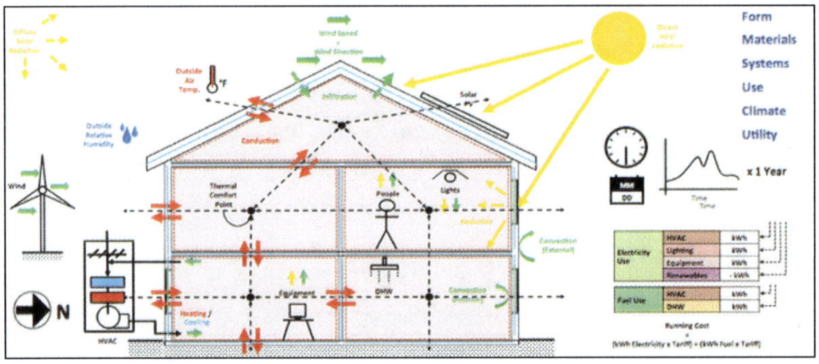

Passive Solar Design

Passive solar design uses sunshine to heat, cool and light homes and other buildings without mechanical or electrical devices. It is usually part of the design of the building itself, using certain materials and placement of windows or skylights.

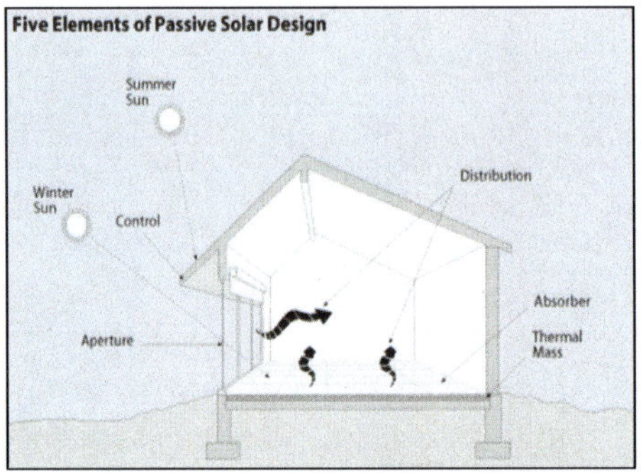

a) Rules of Passive Solar Systems

- The building should be elongated on an east-west axis.
- The building's south face should receive sunlight between the hours of 9:00 A.M. and 3:00 P.M. (sun time) during the heating season.
- Interior spaces requiring the most light and heating and cooling should be along the south face of the building. Less used spaces should be located on the north.

b) The Advantages of Passive Solar Design:

- High energy performance: lower energy bills all year round.
- Investment: independent from future rises in fuel costs, continues to save money long after initial cost recovery.
- Value: high owner satisfaction, high resale value.
- Attractive living environment: large windows and views, sunny interiors, open floor plans.
- Low Maintenance: durable, reduced operation and repair.
- Unwavering comfort: quiet (no operating noise), warmer in winter, cooler in summer (even during a power failure).
- Environmentally friendly: clean, renewable energy doesn't contribute to global warming, acid rain or air pollution.

Passive Solar Heating

The goal of all passive solar heating systems is to capture the sun's heat within the building's elements and release that heat during periods when the sun is not shining. At the same time that the building's elements (or materials) is absorbing heat for later use, solar heat is available for keeping the space comfortable (not overheated).

a) Two primary elements of passive solar heating are required:

- South facing glass.
- Thermal mass to absorb, store, and distribute heat.

b) There are three approaches to passive systems:

- Direct Gain: Sunlight shines into and warms the living space.
- Indirect Gain: Sunlight warms thermal storage, which then warms the living space.
- Isolated Gain: Sunlight warms another room (sunroom) and convection brings the warmed air into the living space.

Water Management in Green Building

Minimising water use is achieved by installing greywater and rainwater catchment systems that recycle water for irrigation or toilet flushing; water-efficient appliances, such as low flow shower-heads, self-closing or spray taps; low-flush toilets, or waterless composting toilets. Installing point of use hot water systems and lagging pipes saves on water heating.

Rainwater Harvesting in Green Building

Rainwater harvesting is the principle of collecting and using precipitation from a catchments surface.

An old technology is gaining popularity in a new way. Rain water harvesting is enjoying a renaissance of sorts in the world, but it traces its history to biblical times.

Extensive rainwater harvesting apparatus existed 4000 years ago in the Palestine and Greece. In ancient Rome, residences were built with individual cisterns and paved courtyards to capture rain water to augment water from city's aqueducts.

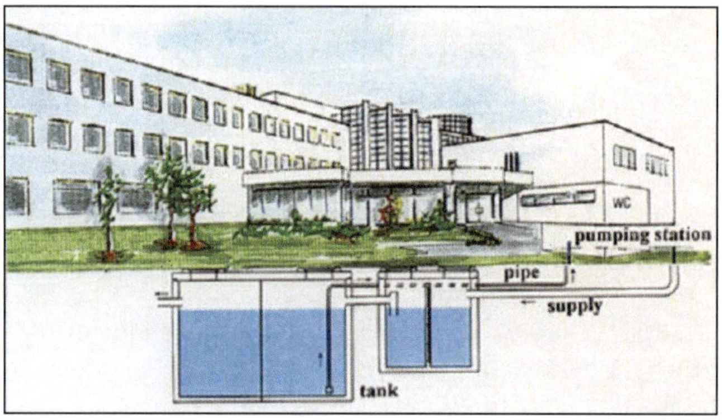

Rainwater Harvesting is Essential

Surface water is inadequate to meet our demand and we have to depend on groundwater. Due to rapid urbanization, infiltration of rainwater into the subsoil has decreased drastically and recharging of groundwater has diminished.

Rainwater Harvesting Techniques for Green Buildings

There are two main techniques of rainwater harvestings:

- Storage of rainwater on surface for future use.
- Recharge to groundwater

Storage of Rainwater on Surface for Future use

The storage of rainwater on surface is a traditional techniques and structures used were underground tanks, ponds, check dams, weirs etc.

Recharge to Groundwater

Recharge to groundwater is a new concept of rainwater harvesting and the structures generally used are Pits, Trenches, Dug wells, Hand pumps, etc.

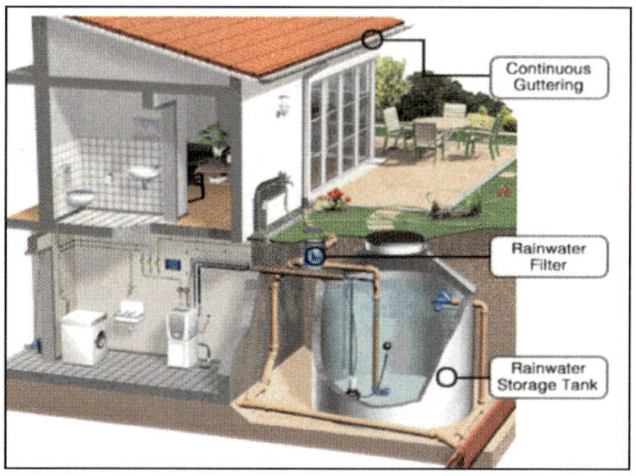

Health Components of Green Building

Using non-toxic materials and products will improve indoor air quality, and reduce the rate of asthma, allergy and sick building syndrome. These materials are emission-free, have low or no VOC content, and are moisture resistant to deter moulds, spores and other microbes.

Indoor air quality is also addressed through ventilation systems and materials that control humidity and allow a building to breathe.

In addition to addressing the above areas, a green building should provide cost savings to the builder and occupants, and meet the broader needs of the community, by using local labour, providing affordable housing, and ensuring the building is sited appropriately for community needs.

Magnetic Concrete Cover Meter Equipment

There are several types of magnetic cover-meter equipment which are used to determine the thickness of reinforcement protector concrete. Commonly, these devices fall into nondestructive tests. Micro-cover meter, Pachometer, Profometer, NOVOTEST, and Cover Master are types of magnetic cover-meter tools.

The devices emit an electromagnetic pulse and detect the magnetic field induced in metal objects. Most cover meters consist of a unit containing the power source, amplifier and meter and a separate search unit (probe) containing the electromagnet which is coupled to the main unit by a cable. Based on the duration or intensity of the induced magnetic field, the location of embedded metal is determined.

The magnetic rebar locator plays a significant role in construction works because the information about the location of steel bars, concrete cover, and bar sizes is essential, directly or indirectly, in many field applications. It is important to note that these magnetic cover-meter devices are not applicable for certain types of bars such as Glass fiber and post-tension or pretension strand.

Types of Magnetic Concrete Cover Meter Equipment

Profometer

It is portable, battery-operated equipment used for measuring the depth of cover concrete, location and size of the steel reinforcement embedded in concrete. The equipment is useful for investigating the structures where drawings are not available. The equipment consists of a data logger, diameter probe and depth probe and calibration block.

The equipment has sufficient memory to store the scanned data. It is used for measuring concrete cover, detecting reinforcing bar, and determining bar size and direction. So, it can be used for corrosion investigation, quality control and assurance of new concrete structures, and location of rebar for penetrations or coring.

Reinforcement cover measurement using profometer.

Pachometer

Pachometer operates by producing a magnetic field and measures the reaction between the magnetic field and the embedded reinforcement. The intensity of the response is associated with the size and location of the steel bars.

The main application of pachometer is to measure the thickness of the concrete cover. So, it can be used to detect and measure protective layer over reinforcing steel bars, prestressed or post-tensioned tendons, wire mesh, and steel pipes.

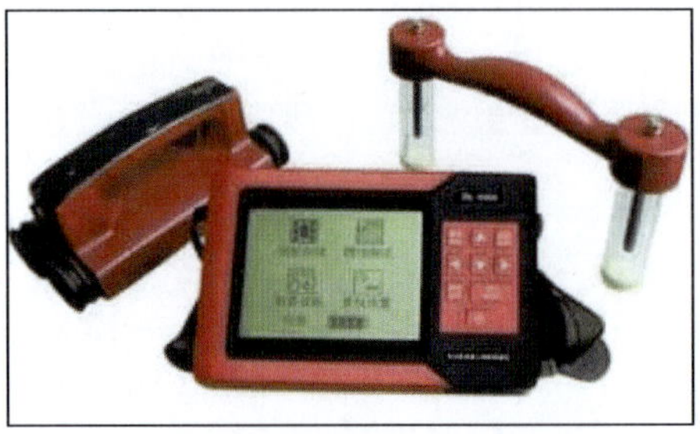

Pachometer concrete cover meter.

Novotest

It is used to measure concrete cover using magnetic method. NOVOTEST is designed according to EN 12504-4, ASTM C597-02, ISO 1920-7:2004, IS13311, CECS 21 standards. It measures cover thickness, determines the diameter of fittings for a known protective layer.

Outstanding advantages of this cover meter device include high precision measurements of concrete cover, possessing a deep search mode, ability to compensate for the influence of parallel rods, and small size and weight

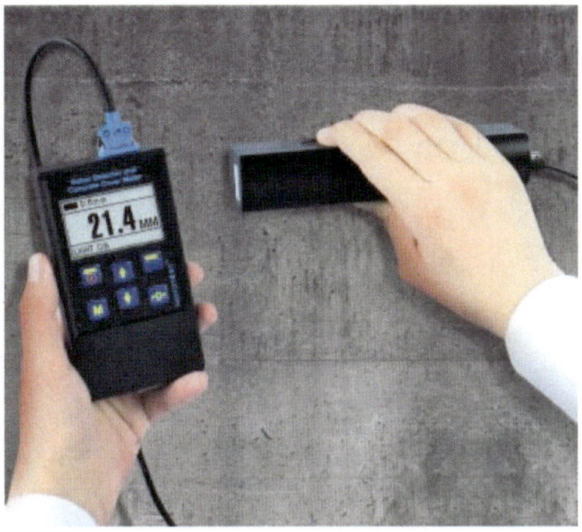

Concrete Cover Meter NOVOTEST Rebar Detector.

Micro Cover Meter

It is easy to use and lightweight cover meter equipment that not only provides rebar location and its size but also concrete protective layer thickness. Micro cover meter is fast, provides clear indication of bar location, and automatically measures bar sizes.

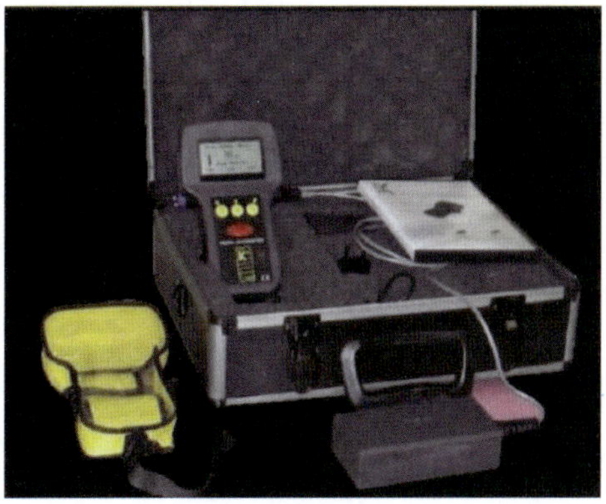

Micro Cover Meter.

Cover Master

Cover master operates easy to use and works automatically; provides concrete cover thickness and steel bar size at the same time. This feature declines the possibility for errors substantially. Cover master yields its result quickly that saves valuable time and money. It is light and easy to handle and highly stable and accurate.

References

- Civil-construction: understandconstruction.com, Retrieved 12 July, 2019
- "Global Status Report 2017 | World Green Building Council". Www.worldgbc.org. Retrieved 2019-03-12
- Commonly-used-building-stones-properties-uses, building-materialm, building: theconstructor.org, Retrieved 11 January, 2019
- Architectural engineer. (n.d.) Mcgraw-Hill Dictionary of Scientific & Technical Terms, 6E. (2003). Retrieved June 8, 2019 from https://encyclopedia2.thefreedictionary.com/Architectural+engineer
- Magnetic-concrete-cover-meter, practical-guide: theconstructor.org, Retrieved 1 August, 2019
- Antunes, Ricardo; Gonzalez, Vicente (3 March 2015). "A Production Model for Construction: A Theoretical Framework". Buildings. 5 (1): 209–228. Doi:10.3390/buildings5010209
- Elements-of-green-building, building: theconstructor.org, Retrieved 5 March, 2019

3

Structural Engineering and Design

Structural engineering is a sub-discipline of civil engineering that focuses on the framework of structures and ensures that they are designed in a way to withstand the stresses and pressures of their environment. It studies various structural materials such as steel frames, reinforced concrete, prestressed concrete, etc. The topics elaborated in this chapter will help in gaining a better perspective about these focus areas of structural engineering.

Structural Engineering is a field of engineering dealing with the analysis and design of structures that support or resist loads. Structural engineering theory is based upon physical laws and empirical knowledge of the structural performance of different materials and geometries. Structural engineering design utilizes a number of simple structural elements to build complex structural systems.

Structural engineering depends upon a detailed knowledge of applied mechanics, materials science and applied mathematics to understand and predict how structures support and resist self-weight and imposed loads. To apply the knowledge successfully a Structural engineering generally requires detailed knowledge of relevant empirical and theoretical design codes, the techniques of structural analysis, as well as some knowledge of the corrosion resistance of the materials and structures, especially when those structures are exposed to the external environment. Since the 1990s, specialist software has become available to aid in the design of structures, with the functionality to assist in the drawing, analyzing and designing of structures with maximum precision; examples include AutoCAD, StaadPro, ETABS, Prokon, Revit Structure etc. Such software may also take into consideration environmental loads, such as from earthquakes and winds.

Structural Design

Structural design is a highly specialized area of civil engineering. It can be described as a set of methods or tools that are used to determine safe and economical specifications for a structure, and to ensure that a planned structure will be sufficiently strong to carry its intended load. Structural engineers conduct a structural analysis to determine what internal and external forces could affect the structure, then design a structure with the appropriate materials and reinforcements to satisfy the requirements.

There is a strong relationship between the professional disciplines of architecture and structural engineering. While architects are normally tasked with designing the visually appealing features of a project, structural engineers place their focus on strength, durability and safety when designing a building.

Basic Principles of Structural Design

Structural engineers combine the core principles of structural design with a sound background in physics and materials science to ensure that structures are built to withstand the loads and forces that they will encounter during their usage.

Civil engineers that design structure for construction projects must be excellent problem solvers. The decisions that structural engineers make during the structural design phase of the project will affect everything from the project cost and duration to the ultimate safety and viability of the structure.

Civil Engineers use Structural Analysis to Understand how Forces Impact Structures

Structural engineers are primarily concerned with the safety and stability of structures that they design. Structural analysis is the main tool that structural engineers use to ensure that a structure can withstand the various types of forces that it will encounter during its normal operation.

A civil engineering designing a subway platform should ensure that the platform will not collapse under the weight of thousands of passengers. A structural engineer designing a railway system should use materials that will not be damaged when a 50,000-lb railway car is used on the tracks.

A structural designer must ensure that the train station they are building won't collapse when snow collects on its rooftops in the winter.

Civil engineers investigate three properties of forces when conducting a structural analysis:

- Magnitude: The size of the force being applied to the structure.
- Direction: The direction of the force. A strong wind would be acting against the side of the building, while the weight of a heavy snowfall would act downwards, in the direction of gravity. Different systems of materials and reinforcements are required to counteract different types of forces.
- Position: The position on which the force acts. Structural engineers must assess the effects of anticipated forces on each area of the building, not just on the building as a whole.

Structural Elements must Remain Stationary

The first step is ensuring through geotechnical engineering that the soil beneath the building provides a solid foundation for construction. A solid foundation helps ensure that the structure itself is immovable and that it can resist external forces.

For a structure to remain stationary, the sum of forces acting on it must always equal zero. However, civil engineers know that external forces will be applied to the structure throughout its operation. These can include live loads such as building occupants and furniture or environmental loads like wind, snow and earthquakes.

In any case, effective structural design requires civil engineers to ensure that a building can absorb applications of external force, maintaining its own equilibrium. Just like a tree branch that doesn't bend will break in the wind, a structure that can't absorb external forces is vulnerable to damage and instability.

Structural Designers must Choose Appropriate Materials

The selection of appropriate building materials is one of the key responsibilities of structural design. The ability of a structure to withstand external and internal forces and maintain its own state of equilibrium and stability depends on the properties of the materials from which it is constructed. Structural engineers choose materials based on a number of measured properties that predict their performance in construction applications:

- Strength, a measure of how well a material can resist an applied stress or load. The opposite of strong is weak.
- Toughness, a measure of how much energy will cause a given material to crack. The opposite of tough is brittle.
- Elasticity is a measure of how much a material can be stretched and still return to its shape.
- Plasticity measures the capacity of a material for plastic deformation. Plastic deformation occurs when a material is stretched beyond its capacity to return to its normal shape. Metals exhibit plastic deformation as they crack, making them tougher than other materials that shatter such as glass.

Steel Frame

Steel frame is a building technique with a "skeleton frame" of vertical steel columns and horizontal I-beams, constructed in a rectangular grid to support the floors, roof and walls of a building which are all attached to the frame. The development of this technique made the construction of the skyscraper possible.

The rolled steel "profile" or cross section of steel columns takes the shape of the letter "I". The two wide flanges of a column are thicker and wider than the flanges on a beam, to better withstand compressive stress in the structure. Square and round tubular sections of steel can also be used, often filled with concrete. Steel beams are connected to the columns with bolts and threaded fasteners, and historically connected by rivets. The central "web" of the steel I-beams is often wider than a column web to resist the higher bending moments that occur in beams.

Wide sheets of steel deck can be used to cover the top of the steel frame as a "form" or corrugated mold, below a thick layer of concrete and steel reinforcing bars. Another popular alternative is a floor of precast concrete flooring units with some form of concrete topping. Often in office buildings, the final floor surface is provided by some form of raised flooring system with the void between the walking surface and the structural floor being used for cables and air handling ducts.

The frame needs to be protected from fire because steel softens at high temperature and this can cause the building to partially collapse. In the case of the columns this is usually done by encasing it in some form of fire resistant structure such as masonry, concrete or plasterboard. The beams may be cased in concrete, plasterboard or sprayed with a coating to insulate it from the heat of the fire or it can be protected by a fire-resistant ceiling construction. Asbestos was a popular material for fireproofing steel structures up until the early 1970s, before the health risks of asbestos fibres were fully understood.

The exterior "skin" of the building is anchored to the frame using a variety of construction techniques and following a huge variety of architectural styles. Bricks, stone, reinforced concrete, architectural glass, sheet metal and simply paint have been used to cover the frame to protect the steel from the weather.

Cold-formed Steel Frames

Interior partition walls made with cold-formed steel.

Cold-formed steel frames are also known as lightweight steel framing (LSF).

Thin sheets of galvanized steel can be cold formed into steel studs for use as a structural or non-structural building material for both external and partition walls in both residential, commercial and industrial construction projects (pictured). The dimension of the room is established with horizontal track that is anchored to the floor and ceiling to outline each room. The vertical studs are arranged in the tracks, usually spaced 16 inches (410 mm) apart, and fastened at the top and bottom.

The typical profiles used in residential construction are the C-shape stud and the U-shaped track, and a variety of other profiles. Framing members are generally produced in a thickness of 12 to 25 gauge. Heavy gauges, such as 12 and 14 gauge, are commonly used when axial loads (parallel to the length of the member) are high, such as in load-bearing construction. Medium-heavy gauges, such as 16 and 18 gauge, are commonly used when there are no axial loads but heavy lateral loads (perpendicular to the member) such as exterior wall studs that need to resist hurricane-force wind loads along coasts. Light gauges, such as 25 gauge, are commonly used where there are no axial loads and very light lateral loads such as in interior construction where the members serve as framing for demising walls between rooms. The wall finish is anchored to the two flange sides of the stud, which varies from $1\frac{1}{4}$ to 3 inches (32 to 76 mm) thick, and the width of web ranges from $1\frac{5}{8}$ to 14 inches (41 to 356 mm). Rectangular sections are removed from the web to provide access for electrical wiring.

Steel mills produce galvanized sheet steel, the base material for the manufacture of cold-formed steel profiles. Sheet steel is then roll-formed into the final profiles used for framing. The sheets are zinc coated (galvanized) to prevent oxidation and corrosion. Steel framing provides excellent design flexibility due to the high strength-to-weight ratio of steel, which allows it to span over a long distances, and also resist wind and earthquake loads.

Steel-framed walls can be designed to offer excellent thermal and acoustic properties – one of the specific considerations when building using cold-formed steel is that thermal bridging can occur across the wall system between the outside environment and interior conditioned space. Thermal bridging can be protected against by installing a layer of externally fixed insulation along the steel framing – typically referred to as a 'thermal break'.

The spacing between studs is typically 16 inches on center for homes exterior and interior walls depending on designed loading requirements. In office suites the spacing is 24 inches (610 mm) on center for all walls except for elevator and staircasewells.

Reinforced Concrete

Reinforced concrete (RC) (also called reinforced cement concrete or RCC) is a composite material in which concrete's relatively low tensile strength and ductility are counteracted by the inclusion of reinforcement having higher tensile strength or ductility. The reinforcement is usually, though not necessarily, steel reinforcing bars (rebar) and is usually embedded passively in the concrete before the concrete sets. Reinforcing schemes are generally designed to resist tensile stresses in particular regions of the concrete that might cause unacceptable cracking and/or structural failure. Modern reinforced concrete can contain varied reinforcing materials made of steel, polymers or alternate composite material in conjunction with rebar or not. Reinforced concrete may also be permanently stressed (concrete in compression, reinforcement in tension), so as to improve the behaviour of the final structure under working loads. In the United States, the most common methods of doing this are known as pre-tensioning and post-tensioning.

For a strong, ductile and durable construction the reinforcement needs to have the following properties at least:

- High relative strength
- High toleration of tensile strain
- Good bond to the concrete, irrespective of pH, moisture, and similar factors
- Thermal compatibility, not causing unacceptable stresses (such as expansion or contraction) in response to changing temperatures.
- Durability in the concrete environment, irrespective of corrosion or sustained stress for example.

Use in Construction

Many different types of structures and components of structures can be built using reinforced concrete including slabs, walls, beams, columns, foundations, frames and more.

Rebars of Sagrada Família's roof in construction.

Reinforced concrete can be classified as precast or cast-in-place concrete. Designing and implementing the most efficient floor system is key to creating optimal building structures. Small changes in the design of a floor system can have significant impact on material costs, construction schedule, ultimate strength, operating costs, occupancy levels and end use of a building. Without reinforcement, constructing modern structures with concrete material would not be possible.

Behavior of Reinforced Concrete

Materials

Concrete is a mixture of coarse (stone or brick chips) and fine (generally sand or crushed stone) aggregates with a paste of binder material (usually Portland cement) and water. When cement is mixed with a small amount of water, it hydrates to form microscopic opaque crystal lattices encapsulating and locking the aggregate into a rigid structure. The aggregates used for making concrete should be free from harmful substances like organic impurities, silt, clay, lignite etc. Typical concrete mixes have high resistance to compressive stresses (about 4,000 psi (28 MPa)); however, any appreciable tension (*e.g.*, due to bending) will break the microscopic rigid lattice, resulting in cracking and separation of the concrete. For this reason, typical non-reinforced concrete must be well supported to prevent the development of tension.

If a material with high strength in tension, such as steel, is placed in concrete, then the composite material, reinforced concrete, resists not only compression but also bending and other direct tensile actions. A composite section where the concrete resists compression and reinforcement "rebar" resists tension can be made into almost any shape and size for the construction industry.

Key Characteristics

Three physical characteristics give reinforced concrete its special properties:

- The coefficient of thermal expansion of concrete is similar to that of steel, eliminating large internal stresses due to differences in thermal expansion or contraction.

- When the cement paste within the concrete hardens, this conforms to the surface details of the steel, permitting any stress to be transmitted efficiently between the different materials. Usually steel bars are roughened or corrugated to further improve the bond or cohesion between the concrete and steel.
- The alkaline chemical environment provided by the alkali reserve (KOH, NaOH) and the portlandite (calcium hydroxide) contained in the hardened cement paste causes a passivating film to form on the surface of the steel, making it much more resistant to corrosion than it would be in neutral or acidic conditions. When the cement paste is exposed to the air and meteoric water reacts with the atmospheric CO_2, portlandite and the calcium silicate hydrate (CSH) of the hardened cement paste become progressively carbonated and the high pH gradually decreases from 13.5 – 12.5 to 8.5, the pH of water in equilibrium with calcite (calcium carbonate) and the steel is no longer passivated.

As a rule of thumb, only to give an idea on orders of magnitude, steel is protected at pH above ~11 but starts to corrode below ~10 depending on steel characteristics and local physico-chemical conditions when concrete becomes carbonated. Carbonation of concrete along with chloride ingress are amongst the chief reasons for the failure of reinforcement bars in concrete.

The relative cross-sectional area of steel required for typical reinforced concrete is usually quite small and varies from 1% for most beams and slabs to 6% for some columns. Reinforcing bars are normally round in cross-section and vary in diameter. Reinforced concrete structures sometimes have provisions such as ventilated hollow cores to control their moisture & humidity.

Distribution of concrete (in spite of reinforcement) strength characteristics along the cross-section of vertical reinforced concrete elements is inhomogeneous.

Mechanism of Composite Action of Reinforcement and Concrete

The reinforcement in a RC structure, such as a steel bar, has to undergo the same strain or deformation as the surrounding concrete in order to prevent discontinuity, slip or separation of the two materials under load. Maintaining composite action requires transfer of load between the concrete and steel. The direct stress is transferred from the concrete to the bar interface so as to change the tensile stress in the reinforcing bar along its length. This load transfer is achieved by means of bond (anchorage) and is idealized as a continuous stress field that develops in the vicinity of the steel-concrete interface.

Anchorage (Bond) in Concrete: Codes of Specifications

Because the actual bond stress varies along the length of a bar anchored in a zone of tension, current international codes of specifications use the concept of development length rather than bond stress. The main requirement for safety against bond failure is to provide a sufficient extension of the length of the bar beyond the point where the steel is required to develop its yield stress and this length must be at least equal to its development length. However, if the actual available length is inadequate for full development, special anchorages must be provided, such as cogs or hooks or mechanical end plates. The same concept applies to lap splice length mentioned in the codes where splices (overlapping) provided between two adjacent bars in order to maintain the required continuity of stress in the splice zone.

Anti-corrosion Measures

In wet and cold climates, reinforced concrete for roads, bridges, parking structures and other structures that may be exposed to deicing salt may benefit from use of corrosion-resistant reinforcement such as uncoated, low carbon/chromium (micro composite), epoxy-coated, hot dip galvanised or stainless steel rebar. Good design and a well-chosen concrete mix will provide additional protection for many applications. Uncoated, low carbon/chromium rebar looks similar to standard carbon steel rebar due to its lack of a coating; its highly corrosion-resistant features are inherent in the steel microstructure. It can be identified by the unique ASTM specified mill marking on its smooth, dark charcoal finish. Epoxy coated rebar can easily be identified by the light green colour of its epoxy coating. Hot dip galvanized rebar may be bright or dull grey depending on length of exposure, and stainless rebar exhibits a typical white metallic sheen that is readily distinguishable from carbon steel reinforcing bar. Reference ASTM standard specifications A1035/A1035M Standard Specification for Deformed and Plain Low-carbon, Chromium, Steel Bars for Concrete Reinforcement, A767 Standard Specification for Hot Dip Galvanised Reinforcing Bars, A775 Standard Specification for Epoxy Coated Steel Reinforcing Bars and A955 Standard Specification for Deformed and Plain Stainless Bars for Concrete Reinforcement.

Another, cheaper way of protecting rebars is coating them with zinc phosphate. Zinc phosphate slowly reacts with calcium cations and the hydroxyl anions present in the cement pore water and forms a stable hydroxyapatite layer.

Penetrating sealants typically must be applied some time after curing. Sealants include paint, plastic foams, films and aluminum foil, felts or fabric mats sealed with tar, and layers of bentonite clay, sometimes used to seal roadbeds. Corrosion inhibitors, such as calcium nitrite [$Ca(NO_2)_2$], can also be added to the water mix before pouring concrete. Generally, 1–2 wt. % of [$Ca(NO_2)_2$] with respect to cement weight is needed to prevent corrosion of the rebars. The nitrite anion is a mild oxidizer that oxidizes the soluble and mobile ferrous ions (Fe^{2+}) present at the surface of the corroding steel and causes them to precipitate as an insoluble ferric hydroxide ($Fe(OH)_3$). This causes the passivation of steel at the anodic oxidation sites. Nitrite is a much more active corrosion inhibitor than nitrate, which is a less powerful oxidizer of the divalent iron.

Reinforcement and Terminology of Beams

Two intersecting beams integral to parking garage slab that will contain both reinforcing steel and the wiring, junction boxes and other electrical components necessary to install the overhead lighting for the garage level beneath it.

A beam bends under bending moment, resulting in a small curvature. At the outer face (tensile face) of the curvature the concrete experiences tensile stress, while at the inner face (compressive face) it experiences compressive stress.

A singly reinforced beam is one in which the concrete element is only reinforced near the tensile face and the reinforcement, called tension steel, is designed to resist the tension.

A doubly reinforced beam is one in which besides the tensile reinforcement the concrete element is also reinforced near the compressive face to help the concrete resist compression. The latter reinforcement is called compression steel. When the compression zone of a concrete is inadequate to resist the compressive moment (positive moment), extra reinforcement has to be provided if the architect limits the dimensions of the section.

An under-reinforced beam is one in which the tension capacity of the tensile reinforcement is smaller than the combined compression capacity of the concrete and the compression steel (under-reinforced at tensile face). When the reinforced concrete element is subject to increasing bending moment, the tension steel yields while the concrete does not reach its ultimate failure condition. As the tension steel yields and stretches, an "under-reinforced" concrete also yields in a ductile manner, exhibiting a large deformation and warning before its ultimate failure. In this case the yield stress of the steel governs the design.

An over-reinforced beam is one in which the tension capacity of the tension steel is greater than the combined compression capacity of the concrete and the compression steel (over-reinforced at tensile face). So the "over-reinforced concrete" beam fails by crushing of the compressive-zone concrete and before the tension zone steel yields, which does not provide any warning before failure as the failure is instantaneous.

A balanced-reinforced beam is one in which both the compressive and tensile zones reach yielding at the same imposed load on the beam, and the concrete will crush and the tensile steel will yield at the same time. This design criterion is however as risky as over-reinforced concrete, because failure is sudden as the concrete crushes at the same time of the tensile steel yields, which gives a very little warning of distress in tension failure.

Steel-reinforced concrete moment-carrying elements should normally be designed to be under-reinforced so that users of the structure will receive warning of impending collapse.

The characteristic strength is the strength of a material where less than 5% of the specimen shows lower strength.

The design strength or nominal strength is the strength of a material, including a material-safety factor. The value of the safety factor generally ranges from 0.75 to 0.85 in Permissible stress design.

The ultimate limit state is the theoretical failure point with a certain probability. It is stated under factored loads and factored resistances.

Reinforced concrete structures are normally designed according to rules and regulations or recommendation of a code such as ACI-318, CEB, Eurocode 2 or the like. WSD, USD or LRFD methods are used in design of RC structural members. Analysis and design of RC members can be carried out by using linear or non-linear approaches. When applying safety factors, building codes normally propose linear approaches, but for some cases non-linear approaches.

Prestressed Concrete

Prestressing concrete is a technique that greatly increases the load-bearing strength of concrete beams. The reinforcing steel in the bottom part of the beam, which will be subjected to tensile forces when in service, is placed in tension before the concrete is poured around it. Once the concrete has hardened, the tension on the reinforcing steel is released, placing a built-in compressive force on the concrete. When loads are applied, the reinforcing steel takes on more stress and the compressive force in the concrete is reduced, but does not become a tensile force. Since the concrete is always under compression, it is less subject to cracking and failure.

Common Failure Modes of Steel Reinforced Concrete

Reinforced concrete can fail due to inadequate strength, leading to mechanical failure, or due to a reduction in its durability. Corrosion and freeze/thaw cycles may damage poorly designed or constructed reinforced concrete. When rebar corrodes, the oxidation products (rust) expand and tends to flake, cracking the concrete and unbonding the rebar from the concrete.

Mechanical Failure

Cracking of the concrete section is nearly impossible to prevent; however, the size and location of cracks can be limited and controlled by appropriate reinforcement, control joints, curing methodology and concrete mix design. Cracking can allow moisture to penetrate and corrode the reinforcement. This is a serviceability failure in limit state design. Cracking is normally the result of an inadequate quantity of rebar, or rebar spaced at too great a distance. The concrete then cracks either under excess loading, or due to internal effects such as early thermal shrinkage while it cures.

Ultimate failure leading to collapse can be caused by crushing the concrete, which occurs when compressive stresses exceed its strength, by yielding or failure of the rebar when bending or shear stresses exceed the strength of the reinforcement, or by bond failure between the concrete and the rebar.

Carbonation

Concrete wall cracking as steel reinforcing corrodes and swells. Rust has a lower density than metal, so it expands as it forms, cracking the decorative cladding off the wall as well as damaging the structural concrete. The breakage of material from a surface is called *spalling*.

Detailed view of spalling probably caused by a too thin layer of concrete between the steel and the surface, accompanied by corrosion from external exposure.

Carbonation, or neutralisation, is a chemical reaction between carbon dioxide in the air and calcium hydroxide and hydrated calcium silicate in the concrete.

When a concrete structure is designed, it is usual to specify the concrete cover for the rebar (the depth of the rebar within the object). The minimum concrete cover is normally regulated by design or building codes. If the reinforcement is too close to the surface, early failure due to corrosion may occur. The concrete cover depth can be measured with a cover meter. However, carbonated concrete incurs a durability problem only when there is also sufficient moisture and oxygen to cause electropotential corrosion of the reinforcing steel.

One method of testing a structure for carbonatation is to drill a fresh hole in the surface and then treat the cut surface with phenolphthalein indicator solution. This solution turns pink when in contact with alkaline concrete, making it possible to see the depth of carbonation. Using an existing hole does not suffice because the exposed surface will already be carbonated.

Chlorides

Chlorides, including sodium chloride, can promote the corrosion of embedded steel rebar if present in sufficiently high concentration. Chloride anions induce both localized corrosion (pitting corrosion) and generalized corrosion of steel reinforcements. For this reason, one should only use fresh raw water or potable water for mixing concrete, ensure that the coarse and fine aggregates do not contain chlorides, rather than admixtures which might contain chlorides.

Rebar for foundations and walls of a sewage pump station.

The Paulins Kill Viaduct, Hainesburg, New Jersey, is 115 feet (35 m) tall and 1,100 feet (335 m) long, and was heralded as the largest reinforced concrete structure in the world when it was completed in 1910 as part of the Lackawanna Cut-Off rail line project. The Lackawanna Railroad was a pioneer in the use of reinforced concrete.

It was once common for calcium chloride to be used as an admixture to promote rapid set-up of the concrete. It was also mistakenly believed that it would prevent freezing. However, this practice fell into disfavor once the deleterious effects of chlorides became known. It should be avoided whenever possible.

The use of de-icing salts on roadways, used to lower the freezing point of water, is probably one of the primary causes of premature failure of reinforced or prestressed concrete bridge decks, roadways, and parking garages. The use of epoxy-coated reinforcing bars and the application of cathodic protection has mitigated this problem to some extent. Also FRP (fiber-reinforced polymer) rebars are known to be less susceptible to chlorides. Properly designed concrete mixtures that have been allowed to cure properly are effectively impervious to the effects of de-icers.

Another important source of chloride ions is sea water. Sea water contains by weight approximately 3.5 wt.% salts. These salts include sodium chloride, magnesium sulfate, calcium sulfate, and bicarbonates. In water these salts dissociate in free ions (Na^+, Mg^{2+}, Cl^-, SO_4^{2-}, HCO_3^-) and migrate with the water into the capillaries of the concrete. Chloride ions, which make up about 50% of these ions, are particularly aggressive as a cause of corrosion of carbon steel reinforcement bars.

In the 1960s and 1970s it was also relatively common for magnesite, a chloride rich carbonate mineral, to be used as a floor-topping material. This was done principally as a levelling and sound attenuating layer. However it is now known that when these materials come into contact with moisture they produce a weak solution of hydrochloric acid due to the presence of chlorides in the magnesite. Over a period of time (typically decades), the solution causes corrosion of the embedded steel rebars. This was most commonly found in wet areas or areas repeatedly exposed to moisture.

Alkali Silica Reaction

This a reaction of amorphous silica (chalcedony, chert, siliceous limestone) sometimes present in the aggregates with the hydroxyl ions (OH^-) from the cement pore solution. Poorly crystallized silica (SiO_2) dissolves and dissociates at high pH (12.5 - 13.5) in alkaline water. The soluble dissociated silicic acid reacts in the porewater with the calcium hydroxide (portlandite) present in the cement paste to form an expansive calcium silicate hydrate (CSH). The alkali–silica reaction (ASR) causes localised swelling responsible for tensile stress and cracking. The conditions required for alkali silica reaction are threefold: (1) aggregate containing an alkali-reactive constituent (amorphous silica), (2) sufficient availability of hydroxyl ions (OH^-), and (3) sufficient moisture, above 75% relative humidity (RH) within the concrete. This phenomenon is sometimes popularly referred to as "concrete cancer". This reaction occurs independently of the presence of rebars; massive concrete structures such as dams can be affected.

Conversion of High Alumina Cement

Resistant to weak acids and especially sulfates, this cement cures quickly and has very high durability and strength. It was frequently used after World War II to make precast concrete objects. However, it can lose strength with heat or time (conversion), especially when not properly cured. After the collapse of three roofs made of prestressed concrete beams using high alumina cement, this cement was banned in the UK in 1976. Subsequent inquiries into the matter showed that the beams were improperly manufactured, but the ban remained.

Sulphates

Sulfates (SO_4) in the soil or in groundwater, in sufficient concentration, can react with the Portland cement in concrete causing the formation of expansive products, e.g., ettringite or thaumasite, which can lead to early failure of the structure. The most typical attack of this type is on concrete slabs and foundation walls at grades where the sulfate ion, via alternate wetting and drying, can increase in concentration. As the concentration increases, the attack on the Portland cement can begin. For buried structures such as pipe, this type of attack is much rarer, especially in the eastern United States. The sulfate ion concentration increases much slower in the soil mass and is especially dependent upon the initial amount of sulfates in the native soil. A chemical analysis of soil borings to check for the presence of sulfates should be undertaken during the design phase of any project involving concrete in contact with the native soil. If the concentrations are found to be aggressive, various protective coatings can be applied. Also, in the US ASTM C150 Type 5 Portland cement can be used in the mix. This type of cement is designed to be particularly resistant to a sulfate attack.

Steel Plate Construction

In steel plate construction, stringers join parallel steel plates. The plate assemblies are fabricated off site, and welded together on-site to form steel walls connected by stringers. The walls become the form into which concrete is poured. Steel plate construction speeds reinforced concrete construction by cutting out the time-consuming on-site manual steps of tying rebar and building forms. The method results in excellent strength because the steel is on the outside, where tensile forces are often greatest.

Fiber-reinforced Concrete

Fiber reinforcement is mainly used in shotcrete, but can also be used in normal concrete. Fiber-reinforced normal concrete is mostly used for on-ground floors and pavements, but can also be considered for a wide range of construction parts (beams, pillars, foundations, etc.), either alone or with hand-tied rebars.

Concrete reinforced with fibers (which are usually steel, glass, plastic fibers) or Cellulose polymer fibre is less expensive than hand-tied rebar. The shape, dimension, and length of the fiber are important. A thin and short fiber, for example short, hair-shaped glass fiber, is only effective during the first hours after pouring the concrete (its function is to reduce cracking while the concrete is stiffening), but it will not increase the concrete tensile strength. A normal-size fiber for European shotcrete (1 mm diameter, 45 mm length—steel or plastic) will increase the concrete's tensile strength. Fiber reinforcement is most often used to supplement or partially replace primary rebar, and in some cases it can be designed to fully replace rebar.

Steel is the strongest commonly available fiber, and comes in different lengths (30 to 80 mm in Europe) and shapes (end-hooks). Steel fibers can only be used on surfaces that can tolerate or avoid corrosion and rust stains. In some cases, a steel-fiber surface is faced with other materials.

Glass fiber is inexpensive and corrosion-proof, but not as ductile as steel. Recently, spun basalt fiber, long available in Eastern Europe, has become available in the U.S. and Western Europe. Basalt fibre is stronger and less expensive than glass, but historically has not resisted the alkaline environment of Portland cement well enough to be used as direct reinforcement. New materials use plastic binders to isolate the basalt fiber from the cement.

The premium fibers are graphite-reinforced plastic fibers, which are nearly as strong as steel, lighter in weight, and corrosion-proof. Some experiments have had promising early results with carbon nanotubes, but the material is still far too expensive for any building.

Non-steel Reinforcement

There is considerable overlap between the subjects of non-steel reinforcement and fiber-reinforcement of concrete. The introduction of non-steel reinforcement of concrete is relatively recent; it takes two major forms: non-metallic rebar rods, and non-steel (usually also non-metallic) fibres incorporated into the cement matrix. For example, there is increasing interest in glass fiber reinforced concrete (GFRC) and in various applications of polymer fibres incorporated into concrete. Although currently there is not much suggestion that such materials will replace metal rebar, some of them have major advantages in specific applications, and there also are new applications in which metal rebar simply is not an option. However, the design and application of non-steel reinforcing is fraught with challenges. For one thing, concrete is a highly alkaline environment, in which many materials, including most kinds of glass, have a poor service life. Also, the behaviour of such reinforcing materials differs from the behaviour of metals, for instance in terms of shear strength, creep and elasticity.

Fibre-reinforced plastic/polymer (FRP) and glass-reinforced plastic (GRP) consist of fibres of polymer, glass, carbon, aramid or other polymers or high-strength fibres set in a resin matrix to form a rebar rod, or grid, or fibres. These rebars are installed in much the same manner as steel rebars. The cost is higher but, suitably applied, the structures have advantages, in particular a dramatic reduction in problems related to corrosion, either by intrinsic concrete alkalinity or by external corrosive fluids that might penetrate the concrete. These structures can be significantly lighter and usually have a longer service life. The cost of these materials has dropped dramatically since their widespread adoption in the aerospace industry and by the military.

In particular, FRP rods are useful for structures where the presence of steel would not be acceptable. For example, MRI machines have huge magnets, and accordingly require non-magnetic buildings. Again, toll booths that read radio tags need reinforced concrete that is transparent to radio waves. Also, where the design life of the concrete structure is more important than its initial costs, non-steel reinforcing often has its advantages where corrosion of reinforcing steel is a major cause of failure. In such situations corrosion-proof reinforcing can extend a structure's life substantially, for example in the intertidal zone. FRP rods may also be useful in situations where it is likely that the concrete structure may be compromised in future years, for example the edges of balconies when balustrades are replaced, and bathroom floors in multi-story construction where the service life of the floor structure is likely to be many times the service life of the waterproofing building membrane.

Plastic reinforcement often is stronger, or at least has a better strength to weight ratio than reinforcing steels. Also, because it resists corrosion, it does not need a protective concrete cover as thick as steel reinforcement does (typically 30 to 50 mm or more). FRP-reinforced structures therefore can be lighter and last longer. Accordingly, for some applications the whole-life cost will be price-competitive with steel-reinforced concrete.

The material properties of FRP or GRP bars differ markedly from steel, so there are differences in the design considerations. FRP or GRP bars have relatively higher tensile strength but lower stiffness, so that deflections are likely to be higher than for equivalent steel-reinforced units. Structures with internal FRP reinforcement typically have an elastic deformability comparable to the plastic deformability (ductility) of steel reinforced structures. Failure in either case is more likely to occur by compression of the concrete than by rupture of the reinforcement. Deflection is always a major design consideration for reinforced concrete. Deflection limits are set to ensure that crack widths in steel-reinforced concrete are controlled to prevent water, air or other aggressive substances reaching the steel and causing corrosion. For FRP-reinforced concrete, aesthetics and possibly water-tightness will be the limiting criteria for crack width control. FRP rods also have relatively lower compressive strengths than steel rebar, and accordingly require different design approaches for reinforced concrete columns.

One drawback to the use of FRP reinforcement is their limited fire resistance. Where fire safety is a consideration, structures employing FRP have to maintain their strength and the anchoring of the forces at temperatures to be expected in the event of fire. For purposes of fireproofing, an adequate thickness of cement concrete cover or protective cladding is necessary. The addition of 1 kg/m^3 of polypropylene fibers to concrete has been shown to reduce spalling during a simulated fire. (The improvement is thought to be due to the formation of pathways out of the bulk of the concrete, allowing steam pressure to dissipate.)

Another problem is the effectiveness of shear reinforcement. FRP rebar stirrups formed by bending before hardening generally perform relatively poorly in comparison to steel stirrups or to structures with straight fibres. When strained, the zone between the straight and curved regions are subject to strong bending, shear, and longitudinal stresses. Special design techniques are necessary to deal with such problems.

There is growing interest in applying external reinforcement to existing structures using advanced materials such as composite (fiberglass, basalt, carbon) rebar, which can impart exceptional strength. Worldwide, there are a number of brands of composite rebar recognized by different countries, such as Aslan, DACOT, V-rod, and ComBar. The number of projects using composite rebar increases day by day around the world, in countries ranging from USA, Russia, and South Korea to Germany.

Prestressed Concrete

Prestressed concrete is a structural material that allows for predetermined, engineering stresses to be placed in members to counteract the stresses that occur when they are subject to loading. It combines the high strength compressive properties of concrete with the high tensile strength of steel.

In ordinary reinforced concrete, stresses are carried by the steel reinforcement, whereas prestressed concrete supports the load by induced stresses throughout the entire structural element. This makes it more resistant to shock and vibration than ordinary concrete, and able to form long, thin structures with much smaller sectional areas to support equivalent loads.

Prestressed concrete was patented by San Franciscan engineer P.H Jackson in 1886, although it did not emerge as an accepted building material until 50 years later when a shortage of steel, coupled with technological advancements, made prestressed concrete the building material of choice during European post-war reconstruction.

It is now commonly used for floor beams, piles and railways sleepers, as well as structures such as bridges, water tanks, roofs and runways. Generally, prestressed concrete is not necessary for columns and walls, however, it can be used economically for tall columns and high retaining walls with high bending stresses.

As a general rule, traditional reinforced concrete is the most economic method for a span of up to 6 m. Prestressed concrete is more economical when spans are over 9 m. Between 6 and 9 m, the two options must be considered according to the particular requirements as to which is the most suitable option.

Steel

Steel used for prestressing may be in the form of wire or tendons that can be grouped to form cables. Solid bars may also be used.

Wire is made by cold-drawing a high carbon steel rod through a series of reducing dies. The wire diameter typically ranges from 3-7 mm and may be round, crimped or indented to give it better bond strength. Another form of tendon is strand which consists of a straight core wire around which is wound in helixes around further wires to give formats such as 7 wire (6 over 1) and 19 wire (9 over 9 over 1). Similar to wire tendons, strand can be used individually or in groups to form cables.

Method

The process of prestressed concrete can be either through pre-tensioning or post-tensioning.

Pre-tensioning

This process involves the stressing of wires or cables by anchoring them at the end of a metal form, which may be up to 120 m in length. Hydraulic jacks stress the wire as required, often adding 10% to accommodate creep and other pre-stress losses that may be incurred. Side moulds are then fixed and the concrete placed around the tensioned wires. The concrete hardens and shrinks, gripping the steel along its length, transferring the tension from the jacks to exert a compressive force in the concrete.

Once the concrete has reached the desired strength, the tensioned wires are released from the jacks. A typical concrete strength of 28 N/mm2 can be achieved by 24-hour steam curing, as well as using additives.

To create shorter members, dividing plates can be placed at any point along the member which, when removed, permit the cutting of the wires.

Post-tensioning

This follows the reverse method to pre-tensioning, whereby the concrete member is cast and the prestressing occurs after the concrete is hardened. This method is often used where stressing is to be carried out on site after casting an insitu component or where a series of precast concrete units are to be joined together to form the required member.

The wires, cables or bars may be positioned in the unit before concreting, but bonding to the concrete is prevented by using a flexible duct or rubber sheath which is deflated and removed when the concrete has hardened.

Stressing is carried out after the concrete has been cured by means of hydraulic jacks operating from one or both ends of the member. Due to the high local stresses at the anchorage positions it is common for a helical (spiral) reinforcement to be included in the design. When the required stress has been reached, the wire or cables are anchored to maintain the prestress. The ends of the unit are sealed with cement mortar to prevent corrosion due to any entrapped moisture and to assist in stress distribution.

Anchorages used in post-tensioning depend on whether the tendons are to be stressed individually or as a group. Most systems use a form of split cone wedges or jaws which act against a form of bearing or pressure plate.

here are many different post-tensioning systems. For example, the Freyssinet system enables the stressing strands to be tensioned simultaneously using centre hole tensioning jacks, anchored by tapered jaws. This is suitable for pre-stressing elements up to 50 m in length.

The Macalloy system on the other hand, involves applying stress to the concrete by means of a solid bar, usually with a diameter of 25-75 mm. The bar is anchored at each end by a special nut which bears against an end plate to distribute the load.

Advantages and Disadvantages

The advantages of prestressed concrete include:

- The inherent compressive strength of concrete is used to its fullest.
- The special alloy steels used to form the prestressing tendons are used to their fullest.
- Tension cracks are eliminated, reducing the risk of the steel components corroding.
- Shear stresses are reduced.
- For any given span and loading condition a reduction in weight can be achieved from using a component with a smaller cross section.
- A composite member can be formed by joining individual precast concrete units together.

The disadvantages of prestressed concrete include:

- A high degree of workmanship and control is required.
- Special alloy steels are more expensive than traditional steels used in reinforced concrete.
- Expensive equipment is needed and there are complex safety requirements.

Timber Framing

Timber framing and “post-and-beam” construction are traditional methods of building with heavy timbers, creating structures using squared-off and carefully fitted and joined timbers with joints secured by large wooden pegs. It is commonplace in wooden buildings through the 19th century. If the structural frame of load-bearing timber is left exposed on the exterior of the building it may be referred to as half-timbered, and in many cases the infill between timbers will be used for decorative effect. The country most known for this kind of architecture is Germany. Timber framed houses are spread all over the country except in the southeast.

The method comes from working directly from logs and trees rather than pre-cut dimensional lumber. Hewing this with broadaxes, adzes, and draw knives and using hand-powered braces and augers (brace and bit) and other woodworking tools, artisans or framers could gradually assemble a building.

Since this building method has been used for thousands of years in many parts of the world, many styles of historic framing have developed. These styles are often categorized by the type of foundation, walls, how and where the beams intersect, the use of curved timbers, and the roof framing details.

Box Frame

A simple timber frame made of straight vertical and horizontal pieces with a common rafter roof without purlins. The term box frame is not well defined and has been used for any kind of framing (with the usual exception of cruck framing). The distinction presented here is that the roof load is carried by the exterior walls. Purlins are also found even in plain timber frames.

Cruck Frame

A "true" or "full" cruck half-timbered building in Weobley, Herefordshire, England: The cruck blades are the tall, curved timbers which extend from near the ground to the ridge.

A cruck is a pair of crooked or curved timbers which form a bent (U.S.) or crossframe (UK); the individual timbers are each called a blade. More than 4,000 cruck frame buildings have been recorded in the UK. Several types of cruck frames are used;

- True cruck or full cruck: Blades, straight or curved, extend from ground or foundation to the ridge acting as the principal rafters. A full cruck does not need a tie beam.
- Base cruck: Tops of the blades are truncated by the first transverse member such as by a tie beam.
- Raised cruck: Blades land on masonry wall, and extend to the ridge.
- Middle cruck: Blades land on masonry wall, and are truncated by a collar.
- Upper cruck: Blades land on a tie beam, very similar to knee rafters.
- Jointed cruck: Blades are made from pieces joined near eaves in a number of ways.
- End cruck is not a style, but on the gable end of a building.

Half-timbered houses, Backnang, Germany.

Half-timbered houses, Miltenberg im Odenwald, Germany.

Rural old railway station timber framing style in Metelen, Germany.

Aisled Frame

Interior of a two-aisled market hall, Chipping Campden, Gloucestershire, England.

Aisled frames have one or more rows of interior posts. These interior posts typically carry more structural load than the posts in the exterior walls. This is the same concept of the aisle in church buildings, sometimes called a hall church, where the center aisle is technically called a nave. However, a nave is often called an aisle, and three-aisled barns are common in the U.S., the Netherlands, and Germany. Aisled buildings are wider than the simpler box-framed or cruck-framed buildings, and typically have purlins supporting the rafters. In northern Germany, this construction is known as variations of a *Ständerhaus*.

Half-timbering

Half-timbered wall with three kinds of infill, wattle and daub, brick, and stone: The plaster coating which originally covered the infill and timbers is mostly gone. This building is in the central German city of Bad Langensalza.

Krämerbrücke in Erfurt, Germany, with half-timbered buildings.

Half-timbering refers to a structure with a frame of load-bearing timber, creating spaces between the timbers called panels, which are then filled-in with some kind of nonstructural material known as infill. The frame is often left exposed on the exterior of the building.

Infill Materials

The earliest known type of infill, called opus craticum by the Romans, was a wattle and daub type construction. Opus craticum is now confusingly applied to a Roman stone/mortar infill as well. Similar methods to wattle and daub were also used and known by various names, such as clam staff and daub, cat-and-clay, or torchis (French), to name only three.

Wattle and daub was the most common infill in ancient times. The sticks were not always technically wattlework (woven), but also individual sticks installed vertically, horizontally, or at an angle into holes or grooves in the framing. The coating of daub has many recipes, but generally was a mixture of clay and chalk with a binder such as grass or straw and water or urine. When the manufacturing of bricks increased, brick infill replaced the less durable infills and became more common. Stone laid in mortar as an infill was used in areas where stone rubble and mortar were available.

Other infills include bousillage, fired brick, unfired brick such as adobe or mudbrick, stones sometimes called pierrotage, planks as in the German ständerbohlenbau, timbers as in ständerblockbau, or rarely cob without any wooden support. The wall surfaces on the interior were often "ceiled" with wainscoting and plastered for warmth and appearance.

Brick infill sometimes called nogging became the standard infill after the manufacturing of bricks made them more available and less expensive. Half-timbered walls may be covered by siding materials including plaster, weatherboarding, tiles, or slate shingles.

The infill may be covered by other materials, including weatherboarding or tiles. or left exposed. When left exposed, both the framing and infill were sometimes done in a decorative manner. Germany is famous for its decorative half-timbering and the figures sometimes have names and meanings. The decorative manner of half-timbering is promoted in Germany by the German Timber-Frame Road, several planned routes people can drive to see notable examples of *Fachwerk* buildings.

The collection of elements in half timbering are sometimes given specific names:

Upper German Fachwerk.

An example of Fachwerk in Franconia.

Cruck framing can be built with half-timber walls. This house is in the Ryedale Folk Museum in England.

Fachwerk in Upper Franconia is very detailed.

Close studding is found in England, Spain and France.

Square-panel half-timbering with fired brick infill: Square paneling is typical of the Low German house, and is found in England.

The term half-timbering is not as old as the German name *Fachwerk* or the French name *colombage*, but it is the standard English name for this style. One of the first people to publish the term "half-timbered" was Mary Martha Sherwood, who employed it in her book, *The Lady of the Manor*, published in several volumes from 1823 to 1829. She uses the term picturesquely: "passing through a gate in a quickset hedge, we arrived at the porch of an old half-timbered cottage, where an aged man and woman received us." This juxtaposition of exposed timbered beams and infilled spaces created the distinctive "half-timbered", or occasionally termed, "Tudor" style, or "black-and-white".

Oldest Examples

The most ancient known half-timbered building is called the House of *opus craticum*. It was buried by the eruption of Mount Vesuvius in 79 AD in Herculaneum, Italy. *Opus craticum* was mentioned by Vitruvius in his books on architecture as a timber frame with wattlework infill. However, the same term is used to describe timber frames with an infill of stone rubble laid in mortar the Romans called *opus incertum*.

Alternative Meanings

A variation of the second meaning of half-timbered: the ground floor is log and the upper floor is framed (half-timbered in the first sense). Kluge House, Montana, U.S.

A less common meaning of the term "half-timbered" is found in the fourth edition of John Henry Parker's Classic Dictionary of Architecture which distinguishes full-timbered houses from half-timbered, with half-timber houses having a ground floor in stone or logs such as the Kluge House which was a log cabin with a timber-framed second floor.

Structure

Joints in an ancient French roof; the wooden pegs hold the mortise and tenon joinery together.

This is a part of a timber frame, before pegs are inserted.

Traditional timber framing is the method of creating framed structures of heavy timber jointed together with various joints, commonly and originally with lap jointing, and then later pegged mortise and tenon joints. Diagonal bracing is used to prevent "racking", or movement of structural vertical beams or posts.

Projecting ("jettied") upper storeys of an English half-timbered village terraced house, the jetties plainly visible.

Originally, German (and other) master carpenters would peg the joints with allowance of about an inch (25 mm), enough room for the wood to move as it 'seasoned', then cut the pegs, and drive the beam home fully into its socket.

To cope with variable sizes and shapes of hewn (by adze or axe) and sawn timbers, two main carpentry methods were employed: scribe carpentry and square rule carpentry.

Scribing or coping was used throughout Europe, especially from the 12th century to the 19th century, and subsequently imported to North America, where it was common into the early 19th century. In a scribe frame, timber sockets are fashioned or "tailor-made" to fit their corresponding timbers; thus, each timber piece must be numbered (or "scribed").

Square-rule carpentry was developed in New England in the 18th century. It used housed joints in main timbers to allow for interchangeable braces and girts. Today, standardized timber sizing means that timber framing can be incorporated into mass-production methods as per the joinery industry, especially where timber is cut by precision computer numerical control machinery.

Jetties

A jetty is an upper floor which sometimes historically used a structural horizontal beam, supported on cantilevers, called a bressummer or 'jetty bressummer' to bear the weight of the new wall, projecting outward from the preceding floor or storey.

In the city of York in the United Kingdom, the famous street known as The Shambles exemplifies this, where jettied houses seem to almost touch above the street.

Timbers

The completed frame of a modern timber-frame house.

Historically, the timbers would have been hewn square using a felling axe and then surface-finished with a broadaxe. If required, smaller timbers were ripsawn from the hewn baulks using pitsaws or frame saws. Today, timbers are more commonly bandsawn, and the timbers may sometimes be machine-planed on all four sides.

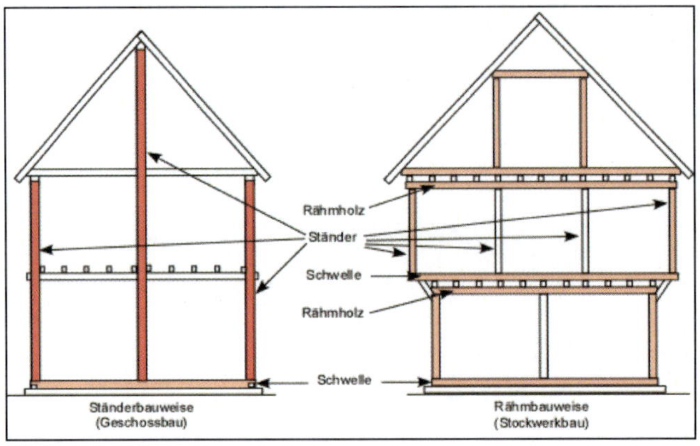

Ridge-post framing (left) and story framing (right, with jetties).

The vertical timbers include:

- Posts (main supports at corners and other major uprights).
- Wall studs (subsidiary upright limbs in framed walls), for example, close studding.

The horizontal timbers include:

- Sill-beams (also called ground-sills or sole-pieces, at the bottom of a wall into which posts and studs are fitted using tenons).
- Noggin-pieces (the horizontal timbers forming the tops and bottoms of the frames of infill panels).
- Wall-plates (at the top of timber-framed walls that support the trusses and joists of the roof).

When jettying, horizontal elements can include:

- The jetty bressummer (or breastsummer), where the main sill (horizontal piece) on which the projecting wall above rests, stretches across the whole width of the jetty wall. The bressummer is itself cantilevered forward, beyond the wall below it.
- The dragon-beam which runs diagonally from one corner to another, and supports the corner posts above and supported by the corner posts below.
- The jetty beams or joists conform t floor dimensions above, but are at right angles to the jetty-plates that conform to the shorter dimensions of "roof" of the floor below. Jetty beams are mortised at 45° into the sides of the dragon beams. They are the main constituents of the cantilever system, and determine how far the jetty projects.
- The jetty-plates are designed to carry the jetty beams. The jetty plates themselves are supported by the corner posts of the recessed floor below.

The sloping timbers include:

- Trusses (the slanting timbers forming the triangular framework at gables and roof).
- Braces (slanting beams giving extra support between horizontal or vertical members of the timber frame).
- Herringbone bracing (a decorative and supporting style of frame, usually at 45° to the upright and horizontal directions of the frame).

Post Construction and Frame Construction

There were two different systems of the position of posts and studs:

- In the older manner, called post construction, the vertical elements continue from the groundwork to the roof. This post construction in German is called *Geschossbauweise* or *Ständerbauweise*. It is somewhat similar to balloon framing method common in North America until the middle of the 20th century.

- In the advanced manner, called frame construction, each story is constructed like a case, and the whole building is constructed like a pile of such cases. This frame construction in German is called *Rähmbauweise* or *Stockwerksbauweise.*

Ridge-post framing is a structurally simple and ancient post and lintel framing where the posts extend all the way to the ridge beams. Germans call this *Firstsäule* or *Hochstud.*

Modern Timber Connector Method

Typical lapped joint assemblies of split-ring connectors.

In the 1930s a system of timber framing referred to as the "modern timber connector method" was developed. It was characterized by the use of timber members assembled into trusses and other framing systems and fastened using various types of metal timber connectors. This type of timber construction was used for various building types including warehouses, factories, garages, barns, stores/markets, recreational buildings, barracks, bridges, and trestles. The use of these structures was promoted because of their low construction costs, easy adaptability, and performance in fire as compared to unprotected steel truss construction.

During World War II, the United States Army Corps of Engineers and the Canadian Military Engineers undertook to construct airplane hangars using this timber construction system in order to conserve steel. Wood hangars were constructed throughout North America and employed various technologies including bowstring, Warren, and Pratt trusses, glued laminated arches, and lamella roof systems. Unique to this building type is the interlocking of the timber members of the roof trusses and supporting columns and their connection points. The timber members are held apart by "fillers" (blocks of timber). This leaves air spaces between the timber members which improves air circulation and drying around the members which improves resistance to moisture born decay.

Shear plate timber connector.

Timber members in this type of framing system were connected with ferrous timber connectors of various types. Loads between timber members were transmitted using split-rings (larger loads), toothed rings (lighter loads), or spiked grid connectors. Split-ring connectors were metal rings sandwiched between adjacent timber members to connect them together. The rings were fit into circular grooves on in both timber members then the assembly was held together with through-bolts. The through-bolts only held the assembly together but were not load-carrying. Shear plate connectors were used to transfer loads between timber members and metal. Shear plate connectors resembled large washers, deformed on the side facing the timber in order to grip it, and were through-fastened with long bolts or lengths of threaded rod. A leading manufacturer of these types of timber connectors was the Timber Engineering Company, or TECO, of Washington, DC. The proprietary name of their split-ring connectors was the "TECO Wedge-Fit."

Modern Features

Interior of a modern hand-hewn post-and-beam home.

Porch of a modern timber-framed house.

A modern *Fachwerk* made by Huf Haus near West Linton, Scotland.

In the United States and Canada, timber-frame construction has been revived since the 1970s, and is now experiencing a thriving renaissance of the ancient skills. This is largely due to such practitioners as Steve Chappell, Jack Sobon, and Tedd Benson, who studied old plans and techniques and revived a long-neglected technique. Once a handcrafted skill passed down, timber-frame construction has now been modernized with the help of modern industrial tools such as CNC machines.These machines and mass-production techniques have assisted growth and made for more affordable frames and shorter lead-times for projects.

Timber-framed structures differ from conventional wood-framed buildings in several ways. Timber framing uses fewer, larger wooden members, commonly timbers in the range of 15 to 30 cm (6 to 12 in), while common wood framing uses many more timbers with dimensions usually in the 5- to 25-cm (2- to 10-in) range. The methods of fastening the frame members also differ. In conventional framing, the members are joined using nails or other mechanical fasteners, whereas timber framing uses the traditional mortise and tenon or more complex joints that are usually fastened using only wooden pegs. Modern complex structures and timber trusses often incorporate steel joinery such as gusset plates, for both structural and architectural purposes.

Recently, it has become common practice to enclose the timber structure entirely in manufactured panels such as structural insulated panels (SIPs). Although the timbers can only be seen from inside the building when so enclosed, construction is less complex and insulation is greater than in traditional timber building. SIPs are "an insulating foam core sandwiched between two structural facings, typically oriented strand board" according to the Structural Insulated Panel Association. SIPs reduce dependency on bracing and auxiliary members, because the panels span considerable distances and add rigidity to the basic timber frame.

An alternate construction method is with concrete flooring with extensive use of glass. This allows a very solid construction combined with open architecture. Some firms have specialized in industrial prefabrication of such residential and light commercial structures such as Huf Haus as low-energy houses or – dependent on location – zero-energy buildings.

Straw-bale construction is another alternative where straw bales are stacked for nonload-bearing infill with various finishes applied to the interior and exterior such as stucco and plaster. This appeals to the traditionalist and the environmentalist as this is using "found" materials to build.

Mudbricks also called adobe are sometimes used to fill in timber-frame structures. They can be made on site and offer exceptional fire resistance. Such buildings must be designed to accommodate the poor thermal insulating properties of mudbrick, however, and usually have deep eaves or a veranda on four sides for weather protection.

Engineered Structures

Timber design or wood design is a subcategory of structural engineering that focuses on the engineering of wood structures. Timber is classified by tree species (e.g., southern pine, douglas fir, etc.) and its strength is graded using numerous coefficients that correspond to the number of knots, the moisture content, the temperature, the grain direction, the number of holes, and other factors. There are design specifications for sawn lumber, glulam members, prefabricated I-joists, composite lumber, and various connection types. In the United States, structural frames are then designed according to the Allowable Stress Design method or the Load Reduced Factor Design method (the latter being preferred).

Advantages

The use of timber framing in buildings offers various aesthetic and structural benefits, as the timber frame lends itself to open plan designs and allows for complete enclosure in effective insulation for energy efficiency. In modern construction, a timber-frame structure offers many benefits:

- It is rapidly erected. A moderately sized timber-frame home can be erected within 2 to 3 days.
- It is well suited to prefabrication, modular construction, and mass-production. Timbers can be pre-fit within bents or wall-sections and aligned with a jig in a shop, without the need for a machine or hand-cut production line. This allows faster erection on site and more precise alignments. Valley and hip timbers are not typically pre-fitted.
- As an alternative to the traditional infill methods, the frame can be encased with SIPs. This stage of preparing the assembled frame for the installation of windows, mechanical systems, and roofing is known as *drying in.*
- It can be customized with carvings or incorporate heirloom structures such as barns etc.

- it can use recycled or otherwise discarded timbers.
- it offers some structural benefits as the timber frame, if properly engineered, lends itself to better *seismic survivability* Consequently, there are many half-timbered houses which still stand despite the foundation having partially caved in over the centuries.
- The generally larger spaces between the frames enable greater flexibility in the placement, at construction or afterwards, of windows and doors with less resulting weakening of the structural integrity and the need for heavy lintels.

In North America, heavy timber construction is classified Building Code Type IV: a special class reserved for timber framing which recognizes the inherent fire resistance of large timber and its ability to retain structural capacity in fire situations. In many cases this classification can eliminate the need and expense of fire sprinklers in public buildings.

Disadvantages

Traditional or Historic Structures

In terms of the traditional half-timber or *fachwerkhaus* there are maybe more disadvantages than advantages today. Such houses are notoriously expensive to maintain let alone renovate and restore, most commonly owing to local regulations that do not allow divergence from the original, modification or incorporation of modern materials. Additionally, in such nations as Germany, where energy efficiency is highly regulated, the renovated building may be required to meet modern energy efficiencies, if it is to be used as a residential or commercial structure (museums and significant historic buildings have no semi-permanent habitade exempt). Many framework houses of significance are treated merely to preserve, rather than render inhabitable – most especially as the required heavy insecticidal fumigation is highly poisonous.

In some cases, it is more economical to build anew using authentic techniques and correct period materials than restore. One major problem with older structures is the phenomenon known as *mechano-sorptive creep* or slanting: where wood beams absorb moisture whilst under compression or tension strains and deform, shift position or both. This is a major structural issue as the house may deviate several degrees from perpendicular to its foundations (in the x-axis, y-axis, and even z-axis) and thus be unsafe and unstable or so out of square it is extremely costly to remedy.

A summary of problems with *Fachwerkhäuser* or half-timbered houses includes the following, though many can be avoided by thoughtful design and application of suitable paints and surface treatments and routine maintenance. Often, though when dealing with a structure of a century or more old, it is too late.

- "Slanting"- *thermo-mechanical* (weather-seasonally induced) and mechano-sorptive (moisture induced) creep of wood in tension and compression.
- Poor prevention of capillary movement of water within any exposed timber, leading to afore-described creep, or rot.
- Eaves that are too narrow or non-existent (thus allowing total exposure to rain and snow).
- Too much exterior detailing that does not allow adequate rainwater run-off.

- Timber ends, joints, and corners poorly protected through coatings, shape or position.
- Non-beveled vertical beams (posts and clapboards) allow water absorption and retention through capillary action.
- Surface point or coatings allowed to deteriorate.
- Traditional gypsum, or wattle and daub containing organic materials (animal hair, straw, manure) which then decompose.
- In both *poteaux-en-terre* and *poteaux-sur-sol* insect, fungus or bacterial decomposition.
- Rot including dry rot.
- Infestation of xylophagous pest organisms such as (very common in Europe) the *Ptinidae* family particularly the common furniture beetle, termites, cockroaches, powderpost beetles, mice, and rats (quite famously so in many children's stories).
- Noise from footsteps in adjacent rooms above, below, and on the same floor in such buildings can be quite audible. This is often resolved with built-up floor systems involving clever sound-isolation and absorption techniques, and at the same time providing passage space for plumbing, wiring, and even heating and cooling equipment.
- Other fungi that are non-destructive to the wood, but are harmful to humans such as black mold. These fungi may also thrive on many "modern" building materials.
- Wood burns more readily than some other materials, making timber-frame buildings somewhat more susceptible to fire damage, although this idea is not universally accepted: Since the cross-sectional dimensions of many structural members exceed 15 cm × 15 cm (6" × 6"), timber-frame structures benefit from the unique properties of large timbers, which char on the outside forming an insulated layer that protects the rest of the beam from burning.
- Prior flood or soil subsidence damage.

Steel Beam Connections

There are various types of steel beam connections used in structures. Steel beam connections are categorized into two groups namely framed and seated connections.

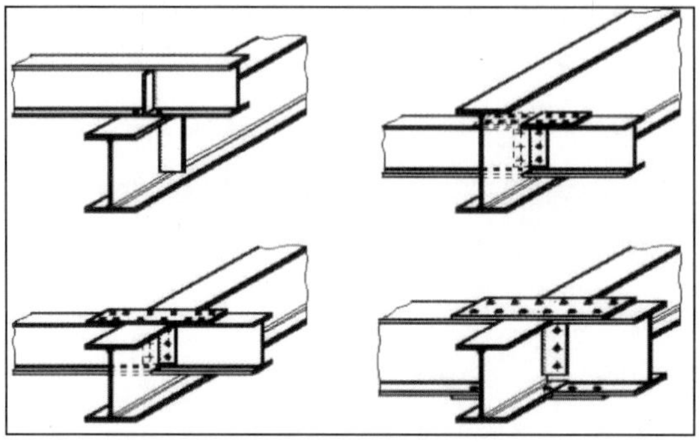

Different Types of Steel Beam to Beam Connections.

In the framed steel beam connections, the beam is connected to the supporting steel element through fittings whereas in case of seated connections, the beam is positioned on seat similar to the case where beam is placed on masonry walls.

Steel Beam to Beam Connections.

Types of Steel Beam Connections

Various types of beam connections are provided below:

- Bolted framed connections,
- Bolted seated connections,
- Welded framed connections,
- Welded seat connections,
- End plate connections,
- Special connections,
- Simple, rigid and semi-rigid connections,

Bolted Framed Steel Beam Connections

In this type of connection, steel beams are linked to supporting elements whether it is steel girders or columns with web connection angle as seen in figure.

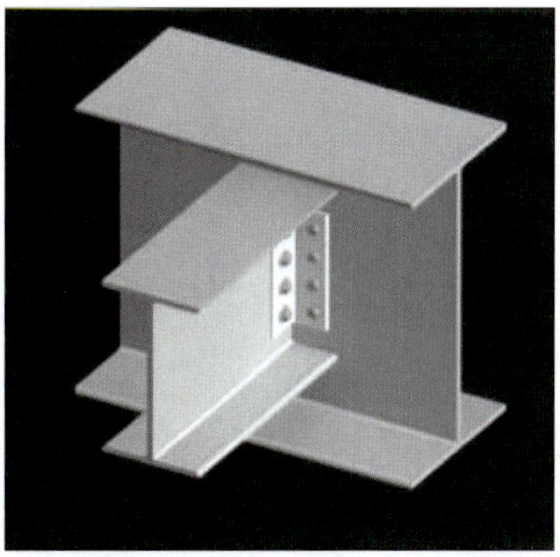

Bolted Framed Steel Beam Connection.

Commonly, the connection is designed based on the loads at the end of the beam. It is required to take strength, type and size of fasteners and strength of base materials into consideration while the connection is designed.

The minimum length of connection angle should be at least half of the beam clear web depth. This measure is specified to ensure sufficient stiffness and stability.

There are various standard sizes of bolted framed connections along with their capacity provided by codes. The purpose of such standardized connection is to increase the speed of the design.

It is recommended to use minimum connection sufficient to withstand the applied load to make the design as economical as possible.

Bolted Seated Steel Beam Connection

There are two major types of bolted seated connections including unstiffened bolted seat connections and stiffened bolted seat connection as illustrated in Figure.

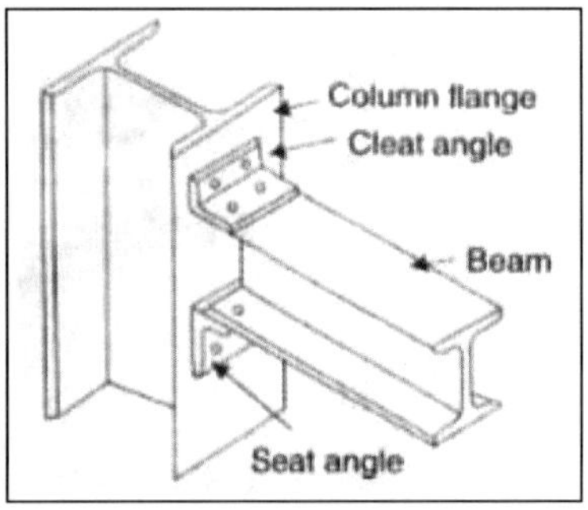

Unstiffened Bolted Seat Connection.

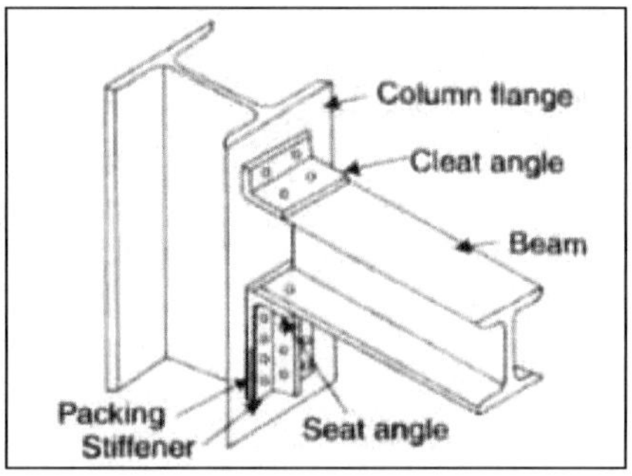

Stiffened Bolted Seat Connection.

If the reactions at the end of the beam is large, then it is recommended to consider stiffened seat connections because it has adequate capacity to resist large forces whereas the capacity of unstiffened seat connection is restricted due to limited bending capacity of seat angle leg that stick out horizontally.

The most outstanding advantage of seat connection is that the beam can be fabricated economically and seat would provide immediate support during erection.

The function of the top angle used in seat connection is to prevent the beam to suffer from accidental rotations.

It is worth mentioning that this type of connection does not need considerable shop detailing. Bolted connections are desired from environmental point of view because the structure can be dismantled and elements can be used in other projects. Added to that, bolted connections can be erected with great ease.

Welded Framed Steel Beam Connection

Like other two types of beam connections, various sizes of welded framed connections with their capacities are available and provided by codes.

The weld of the connection is subjected to direct shear stress and stress caused by loads on the beam that affect weld pattern. So, these stresses are required to be considered.

Clearly, part of the welding is carried out in the field which is challenging since it is difficult to achieve high quality weld due to movements of steel members caused by winds or other factors.

Welded Seat Steel Beam Connections

It is similar to bolt seat connection but welding is used for fastening rather than bolts. The loads on the beam affect the weld pattern eccentrically and generate stress. So, like welded framed connections, such stresses need to be considered.

Types of welded seat connections include unstiffened seat and stiffened seat. The former is used in the case of small applied loads whereas the latter is suitable for case of large loads.

It is advised to use bolts to connect beam bottom flange to the seat. These bolts can be removed or left at their position after the welding process is ended. Welded connection is not desired from environmental point of view and worker. This is because such connection is neither can be dismantled nor erected easily.

End Plate Steel Beam Connection

This type of connection is made possible through the use of welding art. The end plate is connected to the beam web through weld because its capacity and size is governed by shear capacity of the beam web adjoining the weld.

The load applied on the connection at the end of the member does not have eccentricities. There are different types of end plate connections including flexible, semi-rigid, and rigid end plate connections.

It should be known that, fabrications and cuttings should be conducted with utmost care to avoid errors. For example, cutting the beams to length must be as accurate as possible.

Finally, end plate connections are not suitable for tall steel structures.

Special Steel Beam Connections

This type of connections is used for the case where the arrangements of the structural element are in such a way that standard connections cannot be used.

For example, intersection angels might be different by a certain degree and when beam centers offset from column centers.

Examples of special connections include bent-plate framed, single web plate, one sided framed, balanced web plate and Z type connection.

The ability of connections to transfer moments to the columns is based on the degree of fixity of the connections. The higher the degree of fixity of the beam connection, the greater the ability to transfer moments to the columns.

If the connection is designed to transfer moments, then it should be able to withstand shear forces from the beam and transfer moments to the columns.

Special Structures

Rigid Frame Building

The structure of a steel rigid frame building is characterized by its rigid connections of straight or tapered columns and beams. These structures look similar to that of a post and beam system; the difference is its considerable ability to hold vertical loads without additional bracing, due to the rigid connections of its members.

Rigid frame structures were developed in cities like Chicago and Cleveland in the late 19th and early 20th century. This type of frame is preferred for its quick erection, open span, and versatile design use.

Frame Types

Rigid frame, sometimes called continuous frame, offers a rotational stability that enhances how it carries vertical loads, increasing the longevity of the entire structure. Besides rigid frame, the same framing principles are used in the simple frame and partially restrained frame buildings.

The joint connections can be fully fixed-end with fully restrained connections that cannot rotate, or they can be "pin connections," in which the members can freely rotate.

Defining Rigid Frame Construction

In this type of frame, the beam to column connections are classified as rigid, hence the name. The frame is designed to transmit beam end moments and shear forces into the columns without bracing systems to resist lateral loads. The members can be straight or tapered.

The frame stability if only provided by the rigid connections and member stiffness. It looks similar to post and beam but is significantly stronger and able to hold vertical loads.

Local beam-column connection rotations are not considered in global frame analysis; the connections are designed to transmit the resulting beam end movements and shear forces into the columns.

The joints are not always fully fixed to either horizontal or vertical members; when the beam rotates from a vertical load, the columns rotate with it. This allows the joint to rotate as a unit and members maintain the same angular relationships during the rotation.

Rigid frame buildings are highly adaptable and flexible in design. Doors and windows can be placed anywhere, and HVAC units can be placed on the roof or the side. The exterior can easily be dressed to look like any envelope type including stone, brick, or wood.

Differences between Post and Beam and Rigid Frame

- Post and Beam: When a vertical load weighs on a common post and beam structure, it is carried by a horizontal member and then shifted by bending to columns or vertical members. The beam is simply supported by its columns, sitting on top of them so that the ends of the beams can rotate on top of the columns with no restraint. The result is that the horizontal members of the structure then only carries axial forces.
- Rigid Frame: When a rigid frame structure is subjected to a vertical load, it is also picked up by the beams and eventually transferred through the columns to the ground. However, the joints are strongly connected, preventing any free rotation from occurring at the beams ends. This slight difference changes everything about the behavior of the beams, which is now the same as a fix-ended beam.

Rigid frame construction provides many benefits, such as decreased deflections, decreased internal bending moments, and increased rigidity. However, the columns are experiencing some degree of internal bending themselves as the beams stay rigid.

How is it Designed?

The connections of the joints in rigid frame construction vary between being fully fixed-end, in which the connection offers full restraint and no ability to rotate and pin connection, where the members are completely free to rotate. This relationship is referred to the relative stiffness of the horizontal and vertical members.

From a designer's perspective, a rigid frame structure can be designed smaller than other post and beam systems. This is because internal bending moments are reduced by the rigidity. On the other hand, the columns or vertical members should be designed to be a bit larger, since they are carrying both axial loads and internal bending moments.

Moment Resisting Connections

The so-called rigid connections are typically full depth end-plate connections and extended end-plate connections. The most common of these is the bolted end-plate beam-to-column connection. The selection depends on the budget.

Welded connections can be used in place of the bolted end-plate, especially in seismically active regions. Welded connections can provide full moment continuity but tend to be on the expensive side. If used, the welded connections should be prefabricated rather than welded on-site.

Site-bolting takes less time and minimizes labor costs.

Building Applications

Rigid frame is found in a wide variety of building styles and uses:

- Warehouses,
- Retail stores,

- Churches,
- Plants,
- Agricultural buildings,
- Equipment shelters,
- Multi-story buildings of any height.

Rigid frame buildings are typically used when there are special requirements such as medical centers, research facilities, white rooms, and structures housing equipment sensitive to vibrations and deflection.

Benefits of Rigid Frame Construction

Beyond the clear span capabilities, which provide open spaces with no center columns or bracing systems, rigid span steel building retain all the benefits of any metal building.

- Cost-effective,
- Energy-efficient,
- Floors are not sensitive to vibration,
- Connections perform better in load reversal situations and earthquakes.

There are a few disadvantages in that the connections are more complex and can complicate the erection process. The initial cost of the structure is greater as well, but the investment is easily returned by the long life of the building.

Rigid frame construction provides all the benefits of metal buildings and introduces very few drawbacks. Rigid frame allows designers to create large, clear-span spaces that can be used for almost anything. Both the interior and exterior are flexible in appearance while giving you the strength of steel.

Bridge

A bridge is a structure built to span a physical obstacle, such as a body of water, valley, or road, without closing the way underneath. It is constructed for the purpose of providing passage over the obstacle, usually something that can be detrimental to cross otherwise. There are many different designs that each serve a particular purpose and apply to different situations. Designs of bridges vary depending on the function of the bridge, the nature of the terrain where the bridge is constructed and anchored, the material used to make it, and the funds available to build it.

Most likely the earliest bridges were fallen trees and stepping stones, while Neolithic people built boardwalk bridges across marshland. The Arkadiko Bridge dating from the 13th century BC, in the Peloponnese, in southern Greece is one of the oldest arch bridges still in existence and use.

Types of Bridges

Bridges can be categorized in several different ways. Common categories include the type of structural elements used, by what they carry, whether they are fixed or movable, and by the materials used.

Structure Type

Bridges may be classified by how the actions of tension, compression, bending, torsion and shear are distributed through their structure. Most bridges will employ all of these to some degree, but only a few will predominate. The separation of forces and moments may be quite clear. In a suspension or cable-stayed bridge, the elements in tension are distinct in shape and placement. In other cases the forces may be distributed among a large number of members, as in a truss.

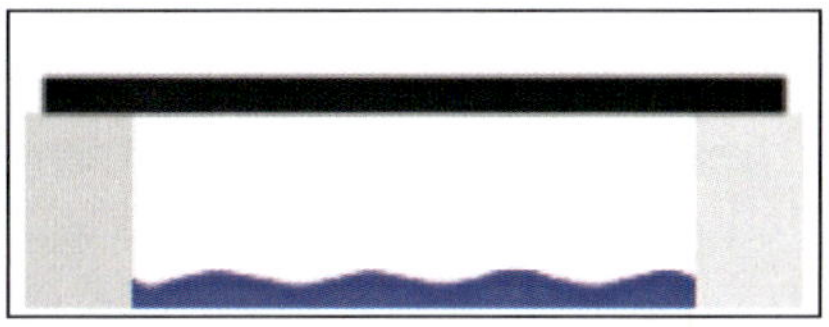

Beam Bridge

Beam bridges are horizontal beams supported at each end by substructure units and can be either *simply supported* when the beams only connect across a single span, or *continuous* when the beams are connected across two or more spans. When there are multiple spans, the intermediate supports are known as piers. The earliest beam bridges were simple logs that sat across streams and similar simple structures. In modern times, beam bridges can range from small, wooden beams to large, steel boxes. The vertical force on the bridge becomes a shear and flexural load on the beam which is transferred down its length to the substructures on either side They are typically made of steel, concrete or wood. Girder bridges and plate girder bridges, usually made from steel, are types of beam bridges. Box girder bridges, made from steel, concrete, or both, are also beam bridges. Beam bridge spans rarely exceed 250 feet (76 m) long, as the flexural stresses increase proportional to the square of the length (and deflection increases proportional to the 4th power of the length). However, the main span of the Rio–Niteroi Bridge, a box girder bridge, is 300 metres (980 ft).

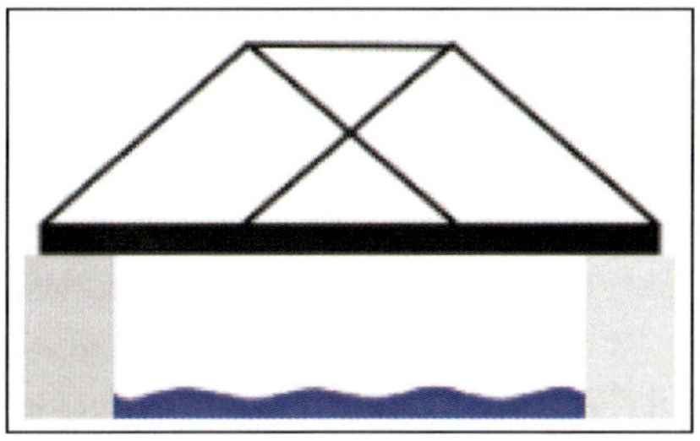

Truss Bridge

The world's longest beam bridge is Lake Pontchartrain Causeway in southern Louisiana in the United States, at 23.83 miles (38.35 km), with individual spans of 56 feet (17 m). Beam bridges are the simplest and oldest type of bridge in use today, and are a popular type.

A truss bridge is a bridge whose load-bearing superstructure is composed of a truss. This truss is a structure of connected elements forming triangular units. The connected elements (typically straight) may be stressed from tension, compression, or sometimes both in response to dynamic loads. Truss bridges are one of the oldest types of modern bridges. The basic types of truss bridges have simple designs which could be easily analyzed by nineteenth and early twentieth century engineers. A truss bridge is economical to construct owing to its efficient use of materials.

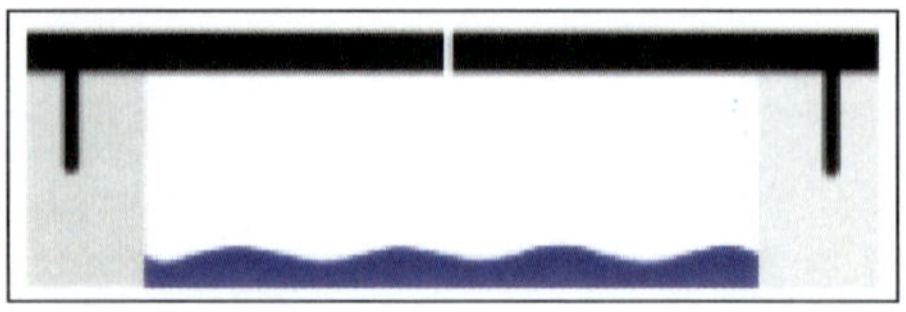

Cantilever Bridge

Cantilever bridges are built using cantilevers—horizontal beams supported on only one end. Most cantilever bridges use a pair of continuous spans that extend from opposite sides of the supporting piers to meet at the center of the obstacle the bridge crosses. Cantilever bridges are constructed using much the same materials and techniques as beam bridges. The difference comes in the action of the forces through the bridge.

Some cantilever bridges also have a smaller beam connecting the two cantilevers, for extra strength. The largest cantilever bridge is the 549-metre (1,801 ft) Quebec Bridge in Quebec, Canada.

Arch bridges have abutments at each end. The weight of the bridge is thrust into the abutments at either side. The earliest known arch bridges were built by the Greeks, and include the Arkadiko Bridge.

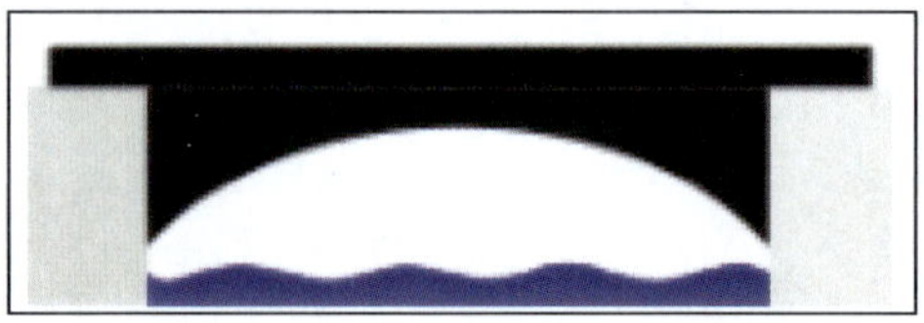

Arch Bridge

With the span of 220 metres (720 ft), the Solkan Bridge over the Soča River at Solkan in Slovenia is the second-largest stone bridge in the world and the longest railroad stone bridge. It was completed in 1905. Its arch, which was constructed from over 5,000 tonnes (4,900 long tons; 5,500 short tons) of stone blocks in just 18 days, is the second-largest stone arch in the world, surpassed only by the Friedensbrücke (Syratalviadukt) in Plauen, and the largest railroad stone arch. The arch of the Friedensbrücke, which was built in the same year, has the span of 90 m (295 ft) and crosses the valley of the Syrabach River. The difference between the two is that the Solkan Bridge was built from stone blocks, whereas the Friedensbrücke was built from a mixture of crushed stone and cement mortar.

The world's largest arch bridge is the Chaotianmen Bridge over the Yangtze River with a length of 1,741 m (5,712 ft) and a span of 552 m (1,811 ft). The bridge was opened April 29, 2009, in Chongqing, China.

Tied Arch Bridge

Tied arch bridges have an arch-shaped superstructure, but differ from conventional arch bridges. Instead of transferring the weight of the bridge and traffic loads into thrust forces into the abutments, the ends of the arches are restrained by tension in the bottom chord of the structure. They are also called bowstring arches.

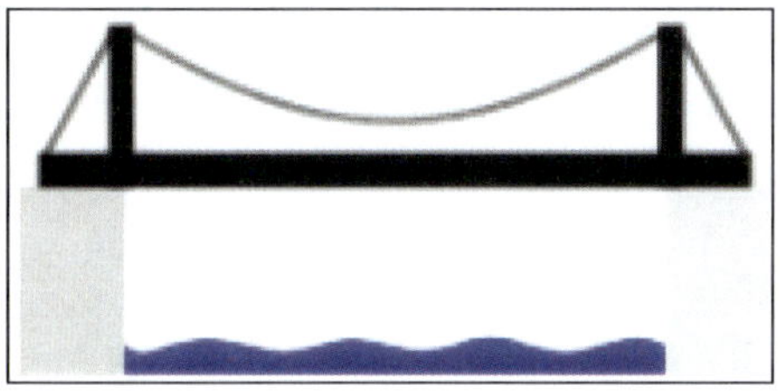

Suspension Bridge

Suspension bridges are suspended from cables. The earliest suspension bridges were made of ropes or vines covered with pieces of bamboo. In modern bridges, the cables hang from towers that are attached to caissons or cofferdams. The caissons or cofferdams are implanted deep into the bed of the lake, river or sea. Sub-types include the simple suspension bridge, the stressed ribbon bridge, the underspanned suspension bridge, the suspended-deck suspension bridge, and the self-anchored suspension bridge. There is also what is sometimes called a "semi-suspension" bridge, of which the Ferry Bridge in Burton-upon-Trent is the only one of its kind in Europe.

The longest suspension bridge in the world is the 3,909 m (12,825 ft) Akashi Kaikyō Bridge in Japan.

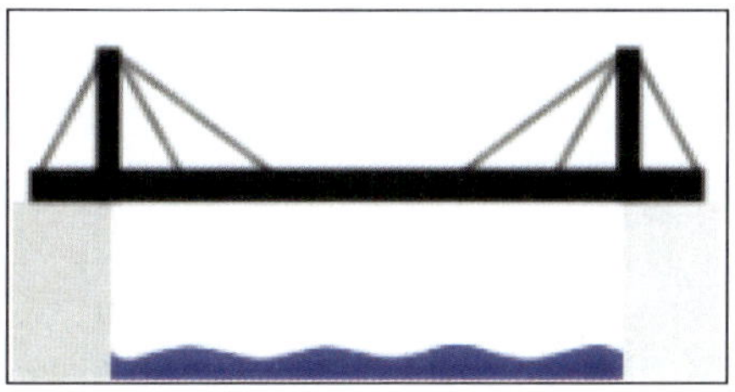

Cable-stayed Bridge

Cable-stayed bridges, like suspension bridges, are held up by cables. However, in a cable-stayed bridge, less cable is required and the towers holding the cables are proportionately higher. The first known cable-stayed bridge was designed in 1784 by C. T. (or C. J.) Löscher.

The longest cable-stayed bridge since 2012 is the 1,104 m (3,622 ft) Russky Bridge in Vladivostok, Russia.

Some Engineers sub-divide 'beam' bridges into slab, beam-and-slab and box girder on the basis of their cross-section. A slab can be solid or voided (though this is no longer favored for inspectability reasons) while beam-and-slab consists of concrete or steel girders connected by a concrete slab. A box-girder cross-section consists of a single-cell or multi-cellular box. In recent years, integral bridge construction has also become popular.

New Champlain Bridge over the St Lawrence river, Montreal, Quebec, Canada.

Fixed or Movable Bridges

Most bridges are fixed bridges, meaning they have no moving parts and stay in one place until they fail or are demolished. Temporary bridges, such as Bailey bridges, are designed to be assembled, and taken apart, transported to a different site, and re-used. They are important in military engineering, and are also used to carry traffic while an old bridge is being rebuilt. Movable bridges are designed to move out of the way of boats or other kinds of traffic, which would otherwise be too tall to fit. These are generally electrically powered.

Double-decked Bridges

The double-decked George Washington Bridge.

Double-decked (or double-decker) bridges have two levels, such as the George Washington Bridge, connecting New York City to Bergen County, New Jersey, US, as the world's busiest bridge, carrying 102 million vehicles annually; truss work between the roadway levels provided stiffness to the roadways and reduced movement of the upper level when the lower level was installed three decades after the upper level. The Tsing Ma Bridge and Kap Shui Mun Bridge in Hong Kong have six lanes on their upper decks, and on their lower decks there are two lanes and a pair of tracks for MTR metro trains. Some double-decked bridges only use one level for street traffic; the Washington Avenue Bridge in Minneapolis reserves its lower level for automobile and light rail traffic and its upper level for pedestrian and bicycle traffic (predominantly students at the University of Minnesota). Likewise, in Toronto, the Prince Edward Viaduct has five lanes of motor traffic, bicycle lanes, and sidewalks on its upper deck; and a pair of tracks for the Bloor–Danforth subway line on its lower deck. The western span of the San Francisco–Oakland Bay Bridge also has two levels.

Robert Stephenson's High Level Bridge across the River Tyne in Newcastle upon Tyne, completed in 1849, is an early example of a double-decked bridge. The upper level carries a railway, and the lower level is used for road traffic. Other examples include Britannia Bridge over the Menai Strait and Craigavon Bridge in Derry, Northern Ireland. The Oresund Bridge between Copenhagen and Malmö consists of a four-lane highway on the upper level and a pair of railway tracks at the lower level. Tower Bridge in London is different example of a double-decked bridge, with the central section consisting of a low-level bascule span and a high-level footbridge.

Viaducts

A viaduct is made up of multiple bridges connected into one longer structure. The longest and some of the highest bridges are viaducts, such as the Lake Pontchartrain Causeway and Millau Viaduct.

Multi-way Bridge

The Tridge is a type of multi-way bridge.

A multi-way bridge has three or more separate spans which meet near the center of the bridge. Multi-way bridges with only three spans appear as a "T" or "Y" when viewed from above. Multi-way bridges bridges are extremely rare. The Tridge, Margaret Bridge, and Zanesville Y-Bridge are examples.

Bridge Types by Use

A bridge can be categorized by what it is designed to carry, such as trains, pedestrian or road traffic (road bridge), a pipeline or waterway for water transport or barge traffic. An aqueduct is a bridge that carries water, resembling a viaduct, which is a bridge that connects points of equal height. A road-rail bridge carries both road and rail traffic. Overway is a term for a bridge that separates incompatible intersecting traffic, especially road and rail. A bridge can carry overhead power lines as does the Storstrøm Bridge.

Some bridges accommodate other purposes, such as the tower of Nový Most Bridge in Bratislava, which features a restaurant, or a bridge-restaurant which is a bridge built to serve as a restaurant. Other suspension bridge towers carry transmission antennas.

Conservationists use wildlife overpasses to stop habitat fragmentation and animal-vehicle collisions. The first animal bridges sprung up in France in the 1950s, and these types of bridges are now used worldwide to protect both large and small wildlife.

Bridges are subject to unplanned uses as well. The areas underneath some bridges have become makeshift shelters and homes to homeless people, and the undertimbers of bridges all around the world are spots of prevalent graffiti. Some bridges attract people attempting suicide, and become known as suicide bridges.

Bridge Types by Material

The Iron Bridge completed in 1781 was the first cast iron bridge.

Krämerbrücke in Erfurt, Germany – with half timbered buildings.

The materials used to build the structure are also used to categorize bridges. Until the end of the 18th century, bridges were made out of timber, stone and masonry. Modern bridges are currently built in concrete, steel, fiber reinforced polymers (FRP), stainless steel or combinations of those materials. Living bridges have been constructed of live plants such as *Ficus elastica* tree roots in India and wisteria vines in Japan.

Small stone bridge, Othonoi, Greece.

Bridge type	Materials used
Cantilever	For small footbridges, the cantilevers may be simple beams; however, large cantilever bridges designed to handle road or rail traffic use trusses built from structural steel, or box girders built from prestressed concrete.
Suspension	The cables are usually made of steel cables galvanised with zinc, along with most of the bridge, but some bridges are still made with steel-reinforced concrete.
Arch	Stone, brick and other such materials that are strong in compression and somewhat so in shear.
Beam	Beam bridges can use pre-stressed concrete, an inexpensive building material, which is then embedded with rebar. The resulting bridge can resist both compression and tension forces.
Truss	The triangular pieces of truss bridges are manufactured from straight and steel bars, according to the truss bridge designs.

Analysis and Design

Unlike buildings whose design is led by architects, bridges are usually designed by engineers. This follows from the importance of the engineering requirements; namely spanning the obstacle and having the durability to survive, with minimal maintenance, in an aggressive outdoor environment. Bridges are first analysed; the bending moment and shear force distributions are calculated due to the applied loads. For this, the finite element method is the most popular. The analysis can be one, two or three-dimensional. For the majority of bridges, a 2-dimensional plate model (often with stiffening beams) is sufficient or an upstand finite element model. On completion of the analysis, the bridge is designed to resist the applied bending moments and shear forces, i.e., section sizes are selected with sufficient capacity to resist the stresses. Many bridges are made of prestressed concrete which has good durability properties, either by pre-tensioning of beams prior to installation or post-tensioning on site.

In most countries, bridges, like other structures, are designed according to Load and Resistance Factor Design (LRFD) principles. In simple terms, this means that the load is factored up by a factor greater than unity, while the resistance or capacity of the structure is factored down, by a factor less than unity. The effect of the factored load (stress, bending moment, etc.) should be less than the factored resistance to that effect. Both of these factors allow for uncertainty and are greater when the uncertainty is greater.

Aesthetics

The Prins Clausbrug across the Amsterdam-Rhine Canal in Utrecht, Netherlands.

The World Heritage Site of Stari Most (Old Bridge) gives its name to the city of Mostar, Bosnia and Herzegovina.

Most bridges are utilitarian in appearance, but in some cases, the appearance of the bridge can have great importance. Often, this is the case with a large bridge that serves as an entrance to a city, or crosses over a main harbor entrance. These are sometimes known as *signature bridges*. Designers of bridges in parks and along parkways often place more importance to aesthetics, as well. Examples include the stone-faced bridges along the Taconic State Parkway in New York.

Bridge at Gatwick Airport. The planes can go underneath the bridge.

To create a beautiful image, some bridges are built much taller than necessary. This type, often found in east-Asian style gardens, is called a Moon bridge, evoking a rising full moon. Other garden bridges may cross only a dry bed of stream washed pebbles, intended only to convey an impression of a stream. Often in palaces a bridge will be built over an artificial waterway as symbolic of a passage to an important place or state of mind. A set of five bridges cross a sinuous waterway in an important courtyard of the Forbidden City in Beijing, China. The central bridge was reserved exclusively for the use of the Emperor, Empress, and their attendants.

Bridge Maintenance

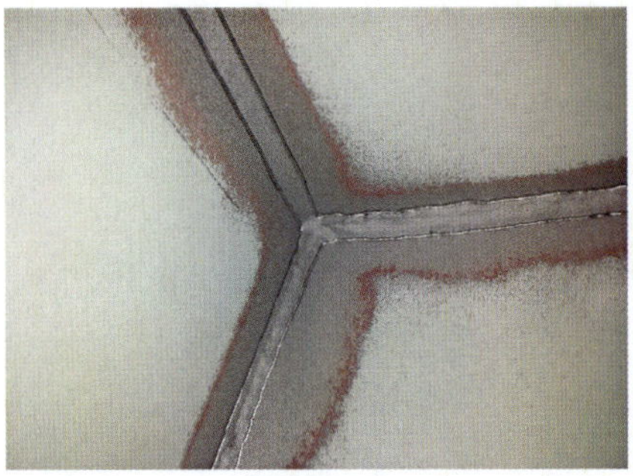

Highway bridge treated with high-frequency impact treatment.

Bridge maintenance consisting of a combination of structural health monitoring and testing. This is regulated in country-specific engineer standards and includes e.g. an ongoing monitoring every three to six months, a simple test or inspection every two to three years and a major inspection every six to ten years. In Europe, the cost of maintenance is considerable and is higher in some countries than spending on new bridges. The lifetime of welded steel bridges can be significantly extended by aftertreatment of the weld transitions. This results in a potential high benefit, using existing bridges far beyond the planned lifetime.

Bridge Traffic Loading

While the response of a bridge to the applied loading is well understood, the applied traffic loading itself is still the subject of research. This is a statistical problem as loading is highly variable, particularly for road bridges. Load Effects in bridges (stresses, bending moments, etc.) are designed for using the principles of Load and Resistance Factor Design. Before factoring to allow for uncertainty, the load effect is generally considered to be the maximum characteristic value in a specified return period. For example, in Europe, it is the maximum value expected in 1000 years.

Bridge standards generally include a load model, deemed to represent the characteristic maximum load to be expected in the return period. In the past, these load models were agreed by standard drafting committees of experts but today, this situation is changing. It is now possible to measure the components of bridge traffic load, i.e., to weigh trucks, using weigh-in-motion (WIM) technologies. With extensive WIM databases, it is possible to calculate the maximum expected load effect in the specified return period. This is an active area of research, addressing issues of opposing direction lanes, side-by-side (same direction) lanes, traffic growth, permit/non-permit vehicles and long-span bridges. Rather than repeat this complex process every time a bridge is to be designed, standards authorities specify simplified notional load models, e.g. HL-93, intended to give the same load effects as the characteristic maximum values. The Eurocode is an example of a standard for bridge traffic loading that was developed in this way.

Traffic Loading on Long Span Bridges

Traffic on Forth Road Bridge, Scotland before it was closed to general traffic. Traffic has now been moved to the Queensferry Crossing, that can be seen on the left.

Most bridge standards are only applicable for short and medium spans - for example, the Eurocode is only applicable for loaded lengths up to 200 m. Longer spans are dealt with on a case by case basis. It is generally accepted that the intensity of load reduces as span increases because the probability of many trucks being closely spaced and extremely heavy reduces as the number of trucks involved increases. It is also generally assumed that short spans are governed by a small number of trucks traveling at high speed, with an allowance for dynamics. Longer spans on the other hand, are governed by congested traffic and no allowance for dynamics is needed. Calculating the loading due to congested traffic remains a challenge as there is a paucity of data on inter-vehicle gaps, both within-lane and inter-lane, in congested conditions. Weigh-in-Motion (WIM) systems provide data on inter-vehicle gaps but only operate well in free flowing traffic conditions. Some authors have used cameras to measure gaps and vehicle lengths in jammed situations and have inferred weights from lengths using WIM data. Others have used microsimulation to generate typical clusters of vehicles on the bridge.

Bridge Vibration

Bridges vibrate under load and this contributes, to a greater or lesser extent, to the stresses. Vibration and dynamics are generally more significant for slender structures such as pedestrian bridges and long-span road or rail bridges. One of the most famous examples is the Tacoma Narrows Bridge that collapsed shortly after being constructed due to excessive vibration. More recently, the Millenium Bridge in London vibrated excessively under pedestrian loading and was closed and retrofitted with a system of dampers. For smaller bridges, dynamics is not catastrophic but can contribute an added amplification to the stresses due to static effects. For example, the Eurocode for bridge loading specifies amplifications of between 10% and 70%, depending on the span, the number of traffic lanes and the type of stress (bending moment or shear force).

Vehicle-bridge Dynamic Interaction

There have been many studies of the dynamic interaction between vehicles and bridges during vehicle crossing events. Fryba did pioneering work on the interaction of a moving load and an Euler-Bernoulli beam. With increased computing power, vehicle-bridge interaction (VBI) models have become ever more sophisticated. The concern is that one of the many natural frequencies associated with the vehicle will resonate with the bridge first natural frequency. The vehicle-related frequencies include body bounce and axle hop but there are also pseudo-frequencies associated with the vehicle's speed of crossing and there are many frequencies associated with the surface profile. Given the wide variety of heavy vehicles on road bridges, a statistical approach has been suggested, with VBI analyses carried out for many statically extreme loading events.

Bridge Failures

The failure of bridges is of special concern for structural engineers in trying to learn lessons vital to bridge design, construction and maintenance. The failure of bridges first assumed national interest during the Victorian era when many new designs were being built, often using new materials.

In the United States, the National Bridge Inventory tracks the structural evaluations of all bridges, including designations such as "structurally deficient" and "functionally obsolete".

Bridge Monitoring

There are several methods used to monitor the condition of large structures like bridges. Many long-span bridges are now routinely monitored with a range of sensors. Many types of sensors are used, including strain transducers, accelerometers, tiltmeters, and GPS. Accelerometers have the advantage that they are inertial, i.e., they do not require a reference point to measure from. This is often a problem for distance or deflection measurement, especially if the bridge is over water.

An option for structural-integrity monitoring is "non-contact monitoring", which uses the Doppler effect (Doppler shift). A laser beam from a Laser Doppler Vibrometer is directed at the point of interest, and the vibration amplitude and frequency are extracted from the Doppler shift of the laser beam frequency due to the motion of the surface. The advantage of this method is that the setup time for the equipment is faster and, unlike an accelerometer, this makes measurements possible on multiple structures in as short a time as possible. Additionally, this method can measure specific points on a bridge that might be difficult to access. However, vibrometers are relatively expensive and have the disadvantage that a reference point is needed to measure from.

Snapshots in time of the external condition of a bridge can be recorded using Lidar to aid bridge inspection. This can provide measurement of the bridge geometry (to facilitate the building of a computer model) but the accuracy is generally insufficient to measure bridge deflections under load.

While larger modern bridges are routinely monitored electronically, smaller bridges are generally inspected visually by trained inspectors. There is considerable research interest in the challenge of smaller bridges as they are often remote and do not have electrical power on site. Possible solutions are the installation of sensors on a specialist inspection vehicle and the use of its measurements as it drives over the bridge to infer information about the bridge condition. These vehicles can be equipped with accelerometers, gyrometers, Laser Doppler Vibrometers and some even have the capability to apply a resonant force to the road surface in order to dynamically excite the bridge at its resonant frequency.

Space Frame

A space frame or space structure (3D truss) is a rigid, lightweight, truss-like structure constructed from interlocking struts in a geometric pattern. Space frames can be used to span large areas with few interior supports. Like the truss, a space frame is strong because of the inherent rigidity of the triangle; flexing loads (bending moments) are transmitted as tension and compression loads along the length of each strut.

Alexander Graham Bell from 1898 to 1908 developed space frames based on tetrahedral geometry. Bell's interest was primarily in using them to make rigid frames for nautical and aeronautical engineering, with the tetrahedral truss being one of his inventions. Dr. Ing. Max Mengeringhausen developed the space grid system called MERO (acronym of MEngeringhausen ROhrbauweise) in 1943 in Germany, thus initiating the use of space trusses in architecture. The commonly used method, still in use has individual tubular members connected at node joints (ball shaped) and variations such as the space deck system, octet truss system and cubic system. Stéphane de Chateau in France invented the Tridirectional SDC system (1957), Unibat system (1959), Pyramitec. A method of tree supports was developed to replace the individual columns. Buckminster Fuller patented the octet truss in 1961 while focusing on architectural structures.

Design Methods

Space frames are typically designed using a rigidity matrix. The special characteristic of the stiffness matrix in an architectural space frame is the independence of the angular factors. If the joints are sufficiently rigid, the angular deflections can be neglected, simplifying the calculations.

Simplified space frame roof with the half-octahedron highlighted in blue.

The simplest form of space frame is a horizontal slab of interlocking square pyramids and tetrahedra built from aluminium or tubular steel struts. In many ways this looks like the horizontal jib of a tower crane repeated many times to make it wider. A stronger form is composed of interlocking tetrahedra in which all the struts have unit length. More technically this is referred to as an isotropic vector matrix or in a single unit width an octet truss. More complex variations change the lengths of the struts to curve the overall structure or may incorporate other geometrical shapes.

Types

Within the meaning of space frame, we can find three systems clearly different between them:

Curvature Classification

- Space plane covers: These spatial structures are composed of planar substructures. Their behavior is similar to that of a plate in which the deflections in the plane are channeled through the horizontal bars and the shear forces are supported by the diagonals.

This train station is supported by a barrel vault structure.

- Barrel vaults: This type of vault has a cross section of a simple arch. Usually this type of space frame does not need to use tetrahedral modules or pyramids as a part of its backing.
- Spherical domes and other compound curves usually require the use of tetrahedral modules or pyramids and additional support from a skin.

Classification by the Arrangement of its Elements

- Single layer grid: All elements are located on the surface to be approximated.
- Double layer grid: Elements are organized in two layers parallel to each other at a certain distance apart. Each of the layers form a lattice of triangles, squares or hexagons in which the projection of the nodes in a layer may overlap or be displaced relative to each other. Diagonal bars connect the nodes of both layers in different directions in space. In this type of meshes, the elements are associated into three groups: upper cordon, cordon and cordon lower diagonal.
- Triple layer grid: Elements are placed in three parallel layers, linked by the diagonals. They are almost always flat.

Other examples classifiable as space frames are these:

- Pleated metallic structures: Emerged to try to solve the problems that formwork and pouring concrete had their counterparts. Typically run with welded joint, but may raise prefabricated joints, a fact which makes them space meshes.
- Hanging covers: Designs on the cable taut, spine, and the catenary arch antifunicular show their ability to channel forces theoretically better than any other alternative, have an infinite range of possibilities for composition and adaptability to any type of plant cover or ensure vain. However, imprecisions in shape having the loaded strand (ideally adapts dynamically to the state of charge) and the risk of bending the arc to unexpected stresses are problems that require pre-compression and prestressing elements. Although in most cases tend to be the cheapest and the technical solution that best fits the acoustics and ventilation of the covered enclosure, are vulnerable to vibration.
- Pneumatic structures: Closure membranes subjected to a pressurized state may be considered within this group.

Applications

Construction

Space frames are a common feature in modern building construction; they are often found in large roof spans in modernist commercial and industrial buildings.

Examples of buildings based on space frames include:

- Stansted Airport, by Foster + Partners.
- Bank of China Tower and the Louvre Pyramid, by I. M. Pei.
- Rogers Centre by Rod Robbie and Michael Allan.
- McCormick Place East in Chicago.
- Eden Project in Cornwall, England.

- Globen, Sweden - Dome with diameter of 110 m.
- Biosphere 2 by John P. Allen, Phil Hawes, Peter Jon Pearce in Oracle, Arizona.
- Jacob K. Javits Convention Center, New York City, New York.
- Sochi International Airport in Sochi, Russia.
- Entrance to Six Flags Magic Mountain.
- Taiwan Taoyuan International Airport airport terminal 2.

Large portable stages and lighting gantries are also frequently built from space frames and octet trusses.

Tunnel

A tunnel is a tube hollowed through soil or stone. Constructing a tunnel, however, is one of the most complex challenges in the field of civil engineering. Many tunnels are considered technological masterpieces and governments have honored tunnel engineers as heroes. That's not to say, of course, that some tunnel projects haven't encountered major setbacks. The Central Artery/Tunnel Project (the "Big Dig") in Boston, Massachusetts was plagued by massive cost overruns, allegations of corruption, and a partial ceiling collapse that resulted in a fatality. But these challenges haven't stopped engineers from dreaming up even bigger and bolder ideas, such as building a Transatlantic Tunnel to connect New York with London.

"Big Dig" in detail to understand the opportunities and challenges inherent to building a tunnel. Finally, we'll look at the future of tunnels.

The Gotthard Base Tunnel, a railway tunnel under construction in Switzerland.

Tunnel Basics

A tunnel is a horizontal passageway located underground. There are many different ways to excavate a tunnel, including manual labor, explosives, rapid heating and cooling, tunneling machinery or a combination of these methods.

Some structures may require excavation similar to tunnel excavation, but are not actually tunnels. Shafts, for example, are often hand-dug or dug with boring equipment. But unlike tunnels, shafts are vertical and shorter. Often, shafts are built either as part of a tunnel project to analyze the rock or soil, or in tunnel construction to provide headings, or locations, from which a tunnel can be excavated.

The diagram below shows the relationship between these underground structures in a typical mountain tunnel. The opening of the tunnel is a portal. The "roof" of the tunnel, or the top half of the tube, is the crown. The bottom half is the invert. The basic geometry of the tunnel is a continuous arch. Because tunnels must withstand tremendous pressure from all sides, the arch is an ideal shape. In the case of a tunnel, the arch simply goes all the way around.

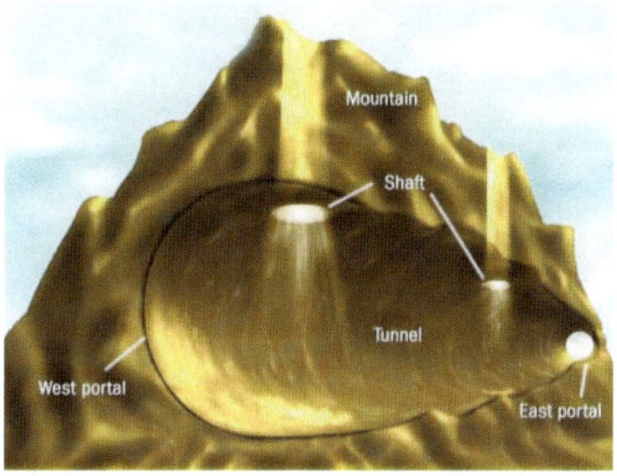

Tunnel engineers, like bridge engineers, must be concerned with an area of physics known as statics. Statics describes how the following forces interact to produce equilibrium on structures such as tunnels and bridges:

- Tension, which expands, or pulls on, material.
- Compression, which shortens, or squeezes material.
- Shearing, which causes parts of a material to slide past one another in opposite directions
- Torsion, which twists a material.

The tunnel must oppose these forces with strong materials, such as masonry, steel, iron and concrete.

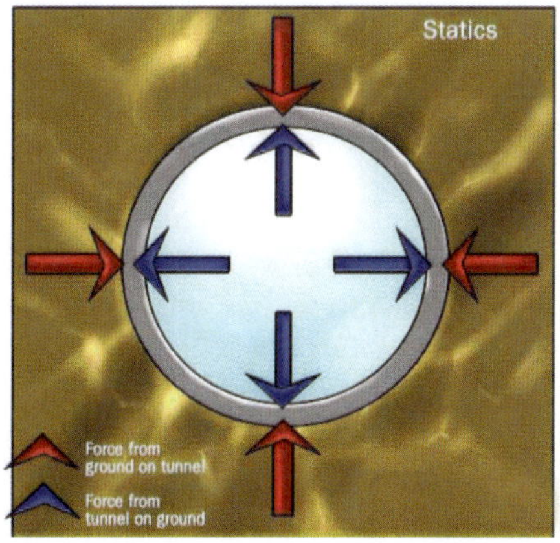

In order to remain static, tunnels must be able to withstand the loads placed on them. Dead load refers to the weight of the structure itself, while live load refers to the weight of the vehicles and people that move through the tunnel.

Types of Tunnels

There are three broad categories of tunnels: mining, public works and transportation. Let's look briefly at each type.

Mine tunnels are used during ore extraction, enabling laborers or equipment to access mineral and metal deposits deep inside the earth. These tunnels are made using similar techniques as other types of tunnels, but they cost less to build. Mine tunnels are not as safe as tunnels designed for permanent occupation, however.

A coal miner standing on the back of a car in a mine tunnel in the early 1900s.
Notice that the sides of the tunnel are shored up with timber.

Public works tunnels carry water, sewage or gas lines across great distances. The earliest tunnels were used to transport water to, and sewage away from, heavily populated regions. Roman engineers used an extensive network of tunnels to help carry water from mountain springs to cities and villages. These tunnels were part of aqueduct systems, which also comprised underground chambers and sloping bridge-like structures supported by a series of arches. By A.D. 97, nine aqueducts carried approximately 85 million gallons of water a day from mountain springs to the city of Rome.

A Roman aqueduct that runs from the Pools of Solomon to Jerusalem.

Before there were trains and cars, there were transportation tunnels such as canals -- artificial waterways used for travel, shipping or irrigation. Just like railways and roadways today, canals usually ran above ground, but many required tunnels to pass efficiently through an obstacle, such as a mountain. Canal construction inspired some of the world's earliest tunnels.

The Underground Canal, located in Lancashire County and Manchester, England, was constructed from the mid- to late-1700s and includes miles of tunnels to house the underground canals. One of America's first tunnels was the Paw Paw Tunnel, built in West Virginia between 1836 and 1850 as part of the Chesapeake and Ohio Canal. Although the canal no longer runs through the Paw Paw, at 3,118 feet long it is still one of the longest canal tunnels in the United States.

Traveling through the Holland Tunnel from Manhattan to New Jersey.

By the 20th century, trains and cars had replaced canals as the primary form of transportation, leading to the construction of bigger, longer tunnels. The Holland Tunnel, completed in 1927, was one of the first roadway tunnels and is still one of the world's greatest engineering projects. Named for the engineer who oversaw construction, the tunnel ushers nearly 100,000 vehicles daily between New York City and New Jersey.

Tunnel Planning

Almost every tunnel is a solution to a specific challenge or problem. In many cases, that challenge is an obstacle that a roadway or railway must bypass. They might be bodies of water, mountains or other transportation routes. Even cities, with little open space available for new construction, can be an obstacle that engineers must tunnel beneath to avoid.

Construction of the Seikan Tunnel involved a 24-year struggle to overcome challenges posed by soft rock under the sea.

In the case of the Holland Tunnel, the challenge was an obsolete ferry system that strained to transport more than 20,000 vehicles a day across the Hudson River. For New York City officials, the solution was clear: Build an automobile tunnel under the river and let commuters drive themselves from New Jersey into the city. The tunnel made an immediate impact. On the opening day alone, 51,694 vehicles made the crossing, with an average trip time of just 8 minutes.

Sometimes, tunnels offer a safer solution than other structures. The Seikan Tunnel in Japan was built because ferries crossing the Tsugaru Strait often encountered dangerous waters and weather conditions. After a typhoon sank five ferryboats in 1954, the Japanese government considered a variety of solutions. They decided that any bridge safe enough to withstand the severe conditions would be too difficult to build. Finally, they proposed a railway tunnel running almost 800 feet below the sea surface. Ten years later, construction began, and in 1988, the Seikan Tunnel officially opened.

How a tunnel is built depends heavily on the material through which it must pass. Tunneling through soft ground, for instance, requires very different techniques than tunneling through hard rock or soft rock, such as shale, chalk or sandstone. Tunneling underwater, the most challenging of all environments, demands a unique approach that would be impossible or impractical to implement above ground.

That's why planning is so important to a successful tunnel project. Engineers conduct a thorough geologic analysis to determine the type of material they will be tunneling through and assess the relative risks of different locations. They consider many factors, but some of the most important include:

- Soil and rock types.
- Weak beds and zones, including faults and shear zones.
- Groundwater, including flow pattern and pressure.
- Special hazards, such as heat, gas and fault lines.

Often, a single tunnel will pass through more than one type of material or encounter multiple hazards. Good planning allows engineers to plan for these variations right from the beginning, decreasing the likelihood of an unexpected delay in the middle of the project.

Once engineers have analyzed the material that the tunnel will pass through and have developed an overall excavation plan, construction can begin. The tunnel engineers' term for building a tunnel is driving, and advancing the passageway can be a long, tedious process that requires blasting, boring and digging by hand.

Tunnel Construction: Soft Ground and Hard Rock

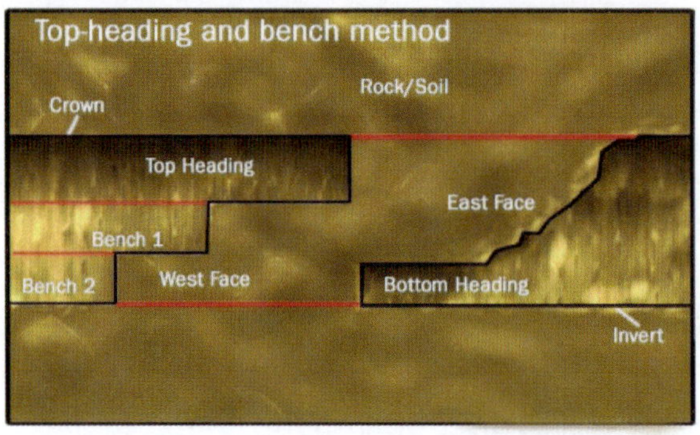

Workers generally use two basic techniques to advance a tunnel. In the full-face method, they excavate the entire diameter of the tunnel at the same time. This is most suitable for tunnels passing through strong ground or for building smaller tunnels. The second technique, shown in the diagram below, is the top-heading-and-bench method. In this technique, workers dig a smaller tunnel known as a heading. Once the top heading has advanced some distance into the rock, workers begin excavating immediately below the floor of the top heading; this is a bench. One advantage of the top-heading-and-bench method is that engineers can use the heading tunnel to gauge the stability of the rock before moving forward with the project.

Notice that the diagram shows tunneling taking place from both sides. Tunnels through mountains or underwater are usually worked from the two opposite ends, or faces, of the passage. In long tunnels, vertical shafts may be dug at intervals to excavate from more than two points. Now let's look more specifically at how tunnels are excavated in each of the four primary environments: soft ground, hard rock, soft rock and underwater.

Soft Ground (Earth)

Workers dig soft-ground tunnels through clay, silt, sand, gravel or mud. In this type of tunnel, stand-up time how long the ground will safely stand by itself at the point of excavation -- is of paramount importance. Because stand-up time is generally short when tunneling through soft ground, cave-ins are a constant threat. To prevent this from happening, engineers use a special piece of equipment called a shield. A shield is an iron or steel cylinder literally pushed into the soft soil. It carves a perfectly round hole and supports the surrounding earth while workers remove debris and install a permanent lining made of cast iron or precast concrete. When the workers complete a section, jacks push the shield forward and they repeat the process.

Marc Isambard Brunel, a French engineer, invented the first tunnel shield in 1825 to excavate the Thames Tunnel in London, England. Brunel's shield comprised 12 connected frames, protected on the top and sides by heavy plates called staves. He divided each frame into three workspaces, or cells, where diggers could work safely. A wall of short timbers, or breasting boards, separated each cell from the face of the tunnel. A digger would remove a breasting board, carve out three or four inches of clay and replace the board. When all of the diggers in all of the cells had completed this process on one section, powerful screw jacks pushed the shield forward.

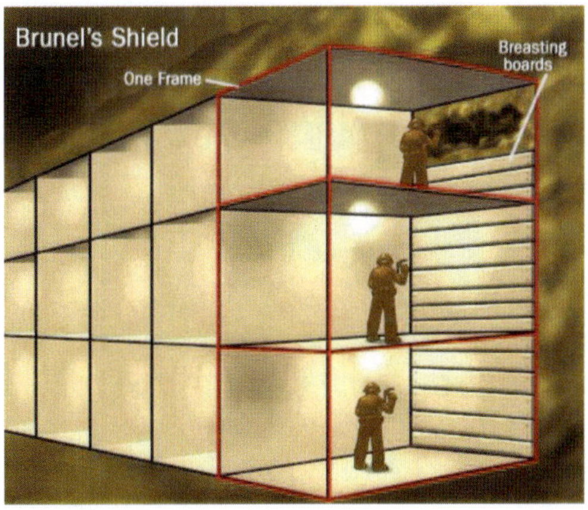

In 1874, Peter M. Barlow and James Henry Greathead improved on Brunel's design by constructing a circular shield lined with cast-iron segments. They first used the newly-designed shield to excavate a second tunnel under the Thames for pedestrian traffic. Then, in 1874, the shield was used to help excavate the London Underground, the world's first subway. Greathead further refined the shield design by adding compressed air pressure inside the tunnel. When air pressure inside the tunnel exceeded water pressure outside, the water stayed out. Soon, engineers in New York, Boston, Budapest and Paris had adopted the Greathead shield to build their own subways.

Hard Rock

Tunneling through hard rock almost always involves blasting. Workers use a scaffold, called a jumbo, to place explosives quickly and safely. The jumbo moves to the face of the tunnel, and drills mounted to the jumbo make several holes in the rock. The depth of the holes can vary depending on the type of rock, but a typical hole is about 10 feet deep and only a few inches in diameter. Next, workers pack explosives into the holes, evacuate the tunnel and detonate the charges. After vacuuming out the noxious fumes created during the explosion, workers can enter and begin carrying out the debris, known as muck, using carts. Then they repeat the process, which advances the tunnel slowly through the rock.

Fire-setting is an alternative to blasting. In this technique, the tunnel wall is heated with fire, and then cooled with water. The rapid expansion and contraction caused by the sudden temperature change causes large chunks of rock to break off. The Cloaca Maxima, one of Rome's oldest sewer tunnels, was built using this technique.

The stand-up time for solid, very hard rock may measure in centuries. In this environment, extra support for the tunnel roof and walls may not be required. However, most tunnels pass through rock that contains breaks or pockets of fractured rock, so engineers must add additional support in the form of bolts, sprayed concrete or rings of steel beams. In most cases, they add a permanent concrete lining.

Tunnel Construction: Soft Rock and Underwater

A TBM boring head showing the disk cutters.

Tunneling through soft rock and tunneling underground require different approaches. Blasting in soft, firm rock such as shale or limestone is difficult to control. Instead, engineers use tunnel-boring machines (TBMs), or moles, to create the tunnel. TBMs are enormous, multimillion-dollar pieces of equipment with a circular plate on one end. The circular plate is covered with disk cutters -- chisel-shaped cutting teeth, steel disks or a combination of the two. As the circular plate slowly rotates, the disk cutters slice into the rock, which falls through spaces in the cutting head onto a conveyor system. The conveyor system carries the muck to the rear of the machine. Hydraulic cylinders attached to the spine of the TBM propel it forward a few feet at a time.

TBMs don't just bore the tunnels they also provide support. As the machine excavates, two drills just behind the cutters bore into the rock. Then workers pump grout into the holes and attach bolts to hold everything in place until the permanent lining can be installed. The TBM accomplishes this with a massive erector arm that raises segments of the tunnel lining into place.

A TBM used in the construction of Yucca Mountain Repository, a U.S. Department of Energy terminal storage facility.

Underwater

Tunnels built across the bottoms of rivers, bays and other bodies of water use the cut-and-cover method, which involves immersing a tube in a trench and covering it with material to keep the tube in place.

Construction begins by dredging a trench in the riverbed or ocean floor. Long, prefabricated tube sections, made of steel or concrete and sealed to keep out water, are floated to the site and sunk in the prepared trench. Then divers connect the sections and remove the seals. Any excess water is pumped out, and the entire tunnel is covered with backfill.

The British end of the Channel Tunnel at Cheriton near Folkestone in Kent.

The tunnel connecting England and France known as the Channel Tunnel, the Euro Tunnel or Chunnel runs beneath the English Channel through 32 miles of soft, chalky earth. Although it's one of the longest tunnels in the world, it took just three years to excavate, thanks to state-of-the-art TBMs. Eleven of these massive machines chewed through the seabed that lay beneath the Channel. Why so many? Because the Chunnel actually consists of three parallel tubes, two that carry trains and one that acts as a service tunnel. Two TBMs placed on opposite ends of the tunnel dug each of these tubes. In essence, the three British TBMs raced against the three French TBMs to see who would make it to the middle first. The remaining five TBMs worked inland, creating the portion of the tunnel that lay between the portals and their respective coasts.

Inside a Holland Tunnel ventilation tower.

Unless the tunnel is short, control of the environment is essential to provide safe working conditions and to ensure the safety of passengers after the tunnel is operational. One of the most important concerns is ventilation a problem magnified by waste gases produced by trains and automobiles. Clifford Holland addressed the problem of ventilation when he designed the tunnel that bears his name. His solution was to add two additional layers above and below the main traffic tunnel. The upper layer clears exhaust fumes, while the lower layer pumps in fresh air. Four large ventilation towers, two on each side of the Hudson River, house the fans that move the air in and out. Eighty-four fans, each 80 feet in diameter, can change the air completely every 90 seconds.

References

- What-is-structural-engineering, resources: civilsimplified.com, Retrieved 8 June, 2019
- Department of Civil Engineering (2015). "History of Concrete Building Construction". CIVL 1101 – History of Concrete. University of Memphis. Retrieved 25 April 2015
- What-is-structural-design-in-civil-engineering: esub.com, Retrieved 3 May, 2019
- Day, Lance (2003). Biographical Dictionary of the History of Technology. Routledge. P. 284. ISBN 0-203-02829-5
- Prestressed-concrete, wik: designingbuildings.co.uk, Retrieved 5 February, 2019
- Kutz, Myer (2011). Handbook of Transportation Engineering, Volume II: Applications and Technologies, Second Edition. Mcgraw-Hill Professional. ISBN 978-0-07-161477-1
- Types-steel-beam-connections, structural-engg: theconstructor.org, Retrieved 30 April, 2019
- "History of Bridges". Historyworld.net. Archived from the original on January 6, 2012. Retrieved January 4, 2012
- What-is-a-rigid-frame-building, bid: whirlwindsteel.com, Retrieved 15 July, 2019
- "Lessons from Roman Cement and Concrete". Pubs.asce.org. Archived from the original on February 10, 2005. Retrieved January 4, 2012
- Tunnel, structural, engineering: howstuffworks.com, Retrieved 30 August, 2019

4
Understanding Geotechnical Engineering

Geotechnical engineering refers to the branch of civil engineering that deals with the application of engineering principles for the acquisition, interpretation and usage of knowledge about the materials of the Earth's crust for solving engineering problems. It uses principles of various fields such as soil mechanics, soil compaction, soil gradation, erosion control, soil nailing, etc. This chapter discusses in detail these principles and theories related to geotechnical engineering.

Geotechnical engineering is the discipline within civil engineering related to the performance of soil mechanics, including their subsurface conditions, determination of the physical, mechanical, and chemical properties that will influence the project under consideration, and the assessment of the risks that may be created by conditions of the site. After necessary evaluation, design of the earthworks is carried out, and subsequently supervision of the site, foundation, and construction is carried out. Geotechnical engineering is also related to coastal engineering, which is involved with the design and creation of harbors, quays and marinas. Today geotechnical engineering has evolved and geosynthetics has helped to solve many conventional problems that are encountered by the engineers in the field. Geosynthetic products have a wide range of applications and are currently used in many civil and geotechnical engineering applications including roads, embankments, retaining structures, canals, landfills, bank protection, and coastal engineering.

One major difference between civil engineering techniques and geotechnical engineering practices is that geotechnical engineering requires competent relevant experiences for sound judgments and relatively less dependency on code based design. A geotechnical engineer's top priority must be to recognize differences in soil and rock properties, evaluate the engineering properties of the rock and soil on the site, and determine the suitable design and construction method, which is simultaneously cost effective, durable, and safe. The scope of work for a geotechnical engineer is huge, and with the invention of new investigative techniques and new materials supporting the work of geotechnical engineer, the field has become even more interesting.

Functions of Geotechnical Engineering

The major functions of geotechnical engineering are as follows:

- Evaluation of geotechnical hazards including potential of landslides.

- Determination of bearing capacity, deformations of foundations, and likely interactions between soil, foundation and the structure.
- Assessment of earth pressure and the performance of retaining walls.
- Analysis of embankment behavior.
- Strength of excavations, caves, tunnels.
- Conducting of response analysis for a site.

Geotechnical Investigation is the first step of applying scientific methods and engineering principles to obtain solutions of civil engineering problems. Identifying soil and rock types, measurement of ground water levels, determining material strength, identifying geological constraints are some of the aspects that a geotechnical engineer needs to study. Moving forward, geotechnical design comes into the picture after investigation business is over and investigation results are taken into due consideration. Feasible performance criterion for engineering structures are laid down through the design process.

Soil Mechanics

Soil mechanics is a branch of soil physics and applied mechanics that describes the behavior of soils. It differs from fluid mechanics and solid mechanics in the sense that soils consist of a heterogeneous mixture of fluids (usually air and water) and particles (usually clay, silt, sand, and gravel) but soil may also contain organic solids and other matter. Along with rock mechanics, soil mechanics provides the theoretical basis for analysis in geotechnical engineering, a subdiscipline of civil engineering, and engineering geology, a subdiscipline of geology. Soil mechanics is used to analyze the deformations of and flow of fluids within natural and man-made structures that are supported on or made of soil, or structures that are buried in soils. Example applications are building and bridge foundations, retaining walls, dams, and buried pipeline systems. Principles of soil mechanics are also used in related disciplines such as engineering geology, geophysical engineering, coastal engineering, agricultural engineering, hydrology and soil physics.

The Leaning Tower of Pisa – an example of a problem due to deformation of soil.

The genesis and composition of soil, the distinction between pore water pressure and inter-granular effective stress, capillary action of fluids in the soil pore spaces, soil classification, seepage and permeability, time dependent change of volume due to squeezing water out of tiny pore spaces, also known as consolidation, shear strength and stiffness of soils. The shear strength of soils is primarily derived from friction between the particles and interlocking, which are very sensitive to the effective stress.

Slope instability issues for a temporary flood control levee in North Dakota, 2009.

Earthwork in Germany.

Fox Glacier, New Zealand: Soil produced and transported by intense weathering and erosion.

Genesis and Composition of Soils

Genesis

The primary mechanism of soil creation is the weathering of rock. All rock types (igneous rock, metamorphic rock and sedimentary rock) may be broken down into small particles to create soil. Weathering mechanisms are physical weathering, chemical weathering, and biological weathering Human activities such as excavation, blasting, and waste disposal, may also create soil. Over geologic time, deeply buried soils may be altered by pressure and temperature to become metamorphic or sedimentary rock, and if melted and solidified again, they would complete the geologic cycle by becoming igneous rock.

Physical weathering includes temperature effects, freeze and thaw of water in cracks, rain, wind, impact and other mechanisms. Chemical weathering includes dissolution of matter composing a rock and precipitation in the form of another mineral. Clay minerals, for example can be formed by weathering of feldspar, which is the most common mineral present in igneous rock.

The most common mineral constituent of silt and sand is quartz, also called silica, which has the chemical name silicon dioxide. The reason that feldspar is most common in rocks but silica is more prevalent in soils is that feldspar is much more soluble than silica.

Silt, Sand, and Gravel are basically little pieces of broken rocks. According to the Unified Soil Classification System, silt particle sizes are in the range of 0.002 mm to 0.075 mm and sand particles have sizes in the range of 0.075 mm to 4.75 mm.

Gravel particles are broken pieces of rock in the size range 4.75 mm to 100 mm. Particles larger than gravel are called cobbles and boulders.

Transport

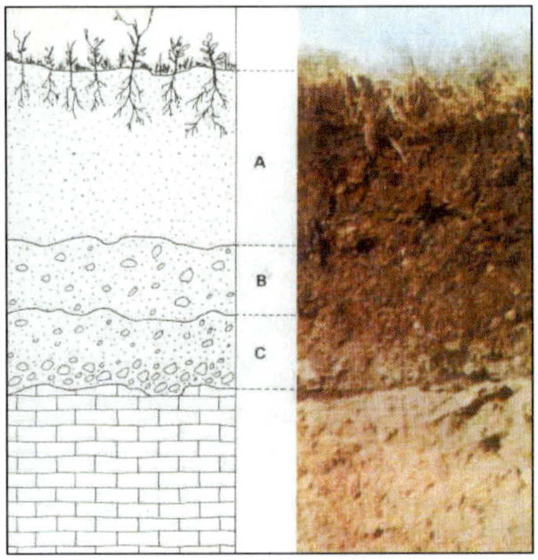

Example soil horizons. a) top soil and colluvium b) mature residual soil c) young residual soil d) weathered rock.

Soil deposits are affected by the mechanism of transport and deposition to their location. Soils that are not transported are called residual soils—they exist at the same location as the rock from which they were generated. Decomposed granite is a common example of a residual soil. The common mechanisms of transport are the actions of gravity, ice, water, and wind. Wind blown soils include dune sands and loess. Water carries particles of different size depending on the speed of the water, thus soils transported by water are graded according to their size. Silt and clay may settle out in a lake, and gravel and sand collect at the bottom of a river bed. Wind blown soil deposits (aeolian soils) also tend to be sorted according to their grain size. Erosion at the base of glaciers is powerful enough to pick up large rocks and boulders as well as soil; soils dropped by melting ice can be a well graded mixture of widely varying particle sizes. Gravity on its own may also carry particles down from the top of a mountain to make a pile of soil and boulders at the base; soil deposits transported by gravity are called colluvium.

The mechanism of transport also has a major effect on the particle shape. For example, low velocity grinding in a river bed will produce rounded particles. Freshly fractured colluvium particles often have a very angular shape.

Soil Composition

Soil Mineralogy

Silts, sands and gravels are classified by their size, and hence they may consist of a variety of minerals. Owing to the stability of quartz compared to other rock minerals, quartz is the most common constituent of sand and silt. Mica, and feldspar are other common minerals present in sands and silts. The mineral constituents of gravel may be more similar to that of the parent rock.

The common clay minerals are montmorillonite or smectite, illite, and kaolinite or kaolin. These minerals tend to form in sheet or plate like structures, with length typically ranging between 10^{-7} m and $4x10^{-6}$ m and thickness typically ranging between 10^{-9} m and $2x10^{-6}$ m, and they have a relatively large specific surface area. The specific surface area (SSA) is defined as the ratio of the surface area of particles to the mass of the particles. Clay minerals typically have specific surface areas in the range of 10 to 1,000 square meters per gram of solid. Due to the large surface area available for chemical, electrostatic, and van der Waals interaction, the mechanical behavior of clay minerals is very sensitive to the amount of pore fluid available and the type and amount of dissolved ions in the pore fluid. To anticipate the effect of clay on the way a soil will behave, it is necessary to know the kinds of clays as well as the amount present. As home builders and highway engineers know all too well, soils containing certain high-activity clays make very unstable material on which to build because they swell when wet and shrink when dry. This shrink-and-swell action can easily crack foundations and cause retaining walls to collapse. These clays also become extremely sticky and difficult to work with when they are wet. In contrast, low-activity clays, formed under different conditions, can be very stable and easy to work with.

The minerals of soils are predominantly formed by atoms of oxygen, silicon, hydrogen, and aluminum, organized in various crystalline forms. These elements along with calcium, sodium, potassium, magnesium, and carbon constitute over 99 percent of the solid mass of soils.

Grain Size Distribution

Soils consist of a mixture of particles of different size, shape and mineralogy. Because the size of the particles obviously has a significant effect on the soil behavior, the grain size and grain size distribution are used to classify soils. The grain size distribution describes the relative proportions of particles of various sizes. The grain size is often visualized in a cumulative distribution graph which, for example, plots the percentage of particles finer than a given size as a function of size. The median grain size, D_{50}, is the size for which 50% of the particle mass consists of finer particles.

Soil behavior, especially the hydraulic conductivity, tends to be dominated by the smaller particles, hence, the term "effective size", denoted by D_{10}, is defined as the size for which 10% of the particle mass consists of finer particles.

Sands and gravels that possess a wide range of particle sizes with a smooth distribution of particle sizes are called well graded soils. If the soil particles in a sample are predominantly in a relatively narrow range of sizes, the sample is uniformly graded. If a soil sample has distinct gaps in the gradation curve, e.g., a mixture of gravel and fine sand, with no coarse sand, the sample may be gap graded. Uniformly graded and gap graded soils are both considered to be poorly graded. There are many methods for measuring particle-size distribution. The two traditional methods are sieve analysis and hydrometer analysis.

Sieve Analysis

Sieve.

The size distribution of gravel and sand particles are typically measured using sieve analysis. The formal procedure is described in ASTM D6913-04. A stack of sieves with accurately dimensioned holes between a mesh of wires is used to separate the particles into size bins. A known volume of dried soil, with clods broken down to individual particles, is put into the top of a stack of sieves arranged from coarse to fine. The stack of sieves is shaken for a standard period of time so that the particles are sorted into size bins. This method works reasonably well for particles in the sand and gravel size range. Fine particles tend to stick to each other, and hence the sieving process is not an effective method. If there are a lot of fines (silt and clay) present in the soil it may be necessary to run water through the sieves to wash the coarse particles and clods through.

A variety of sieve sizes are available. The boundary between sand and silt is arbitrary. According to the Unified Soil Classification System, a #4 sieve (4 openings per inch) having 4.75mm opening size separates sand from gravel and a #200 sieve with an 0.075 mm opening separates sand from silt and clay. According to the British standard, 0.063 mm is the boundary between sand and silt, and 2 mm is the boundary between sand and gravel.

Hydrometer Analysis

The classification of fine-grained soils, i.e., soils that are finer than sand, is determined primarily by their Atterberg limits, not by their grain size. If it is important to determine the grain size distribution of fine-grained soils, the hydrometer test may be performed. In the hydrometer tests, the soil particles are mixed with water and shaken to produce a dilute suspension in a glass cylinder, and then the cylinder is left to sit. A hydrometer is used to measure the density of the suspension as a function of time. Clay particles may take several hours to settle past the depth of measurement of the hydrometer. Sand particles may take less than a second. Stoke's law provides the theoretical basis to calculate the relationship between sedimentation velocity and particle size. ASTM provides the detailed procedures for performing the Hydrometer test.

Clay particles can be sufficiently small that they never settle because they are kept in suspension by Brownian motion, in which case they may be classified as colloids.

Mass-volume Relations

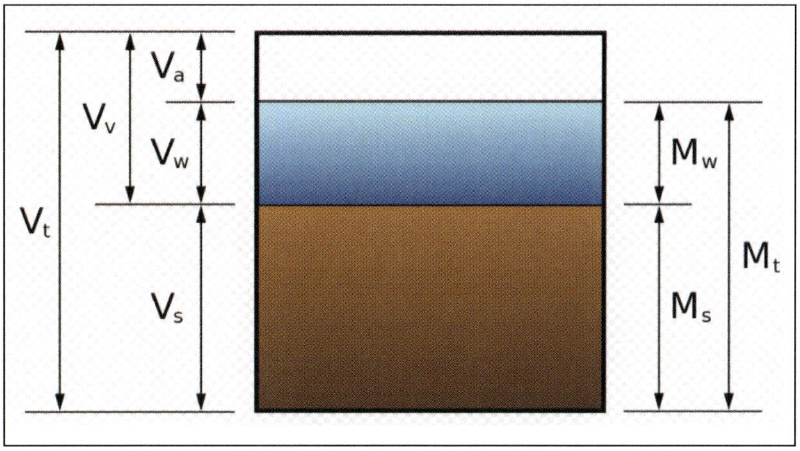

A phase diagram of soil indicating the masses and volumes of air, solid, water, and voids.

There are a variety of parameters used to describe the relative proportions of air, water and solid in a soil. The basic notation is as follows:

V_a, V_w, and V_s represent the volumes of air, water and solids in a soil mixture;

W_a, W_w, and W_s represent the weights of air, water and solids in a soil mixture;

M_a, M_w, and M_s represent the masses of air, water and solids in a soil mixture;

ρ_a, ρ_w, and ρ_s represent the densities of the constituents (air, water and solids) in a soil mixture;

Note that the weights, W, can be obtained by multiplying the mass, M, by the acceleration due to gravity, g; e.g., $W_s = M_s g$

Specific Gravity is the ratio of the density of one material compared to the density of pure water ($\rho_w = 1g/cm^3$).

Specific gravity of solids, $G_s = \frac{\rho_s}{\rho_w}$

Note that specific weight, conventionally denoted by the symbol γ may be obtained by multiplying the density (ρ) of a material by the acceleration due to gravity, g.

Density, Bulk Density, or Wet Density, ρ, are different names for the density of the mixture, i.e., the total mass of air, water, solids divided by the total volume of air water and solids (the mass of air is assumed to be zero for practical purposes):

$$\rho = \frac{M_s + M_w}{V_s + V_w + V_a} = \frac{M_t}{V_t}$$

Dry Density, ρ_d, is the mass of solids divided by the total volume of air water and solids:

$$\rho_d = \frac{M_s}{V_s + V_w + V_a} = \frac{M_s}{V_t}$$

Buoyant Density, ρ', defined as the density of the mixture minus the density of water is useful if the soil is submerged under water:

$$\rho' = \rho - \rho_w$$

where ρ_w is the density of water

Water Content, is the ratio of mass of water to mass of solid. It is easily measured by weighing a sample of the soil, drying it out in an oven and re-weighing. Standard procedures are described by ASTM.

$$w = \frac{M_w}{M_s} = \frac{W_w}{W_s}$$

Void ratio, e, is the ratio of the volume of voids to the volume of solids:

$$e = \frac{V_v}{V_s} = \frac{V_v}{V_T - V_V} = \frac{n}{1-n}$$

Porosity, n, is the ratio of volume of voids to the total volume, and is related to the void ratio:

$$n = \frac{V_v}{V_t} = \frac{V_v}{V_s + V_v} = \frac{e}{1+e}$$

Degree of saturation, S, is the ratio of the volume of water to the volume of voids:

$$S = \frac{V_w}{V_v}$$

From the above definitions, some useful relationships can be derived by use of basic algebra.

$$\rho = \frac{(G_s + Se)\rho_w}{1+e}$$

$$\rho = \frac{(1+w)G_s\rho_w}{1+e}$$

$$w = \frac{Se}{G_s}$$

Soil Classification

Geotechnical engineers classify the soil particle types by performing tests on disturbed (dried, passed through sieves, and remolded) samples of the soil. This provides information about the characteristics of the soil grains themselves. Classification of the types of grains present in a soil does not account for important effects of the *structure* or *fabric* of the soil, terms that describe compactness of the particles and patterns in the arrangement of particles in a load carrying framework as well as the pore size and porc fluid distributions. Engineering geologists also classify soils based on their genesis and depositional history.

Classification of Soil Grains

In the US and other countries, the Unified Soil Classification System (USCS) is often used for soil classification. Other classification systems include the British Standard BS 5930 and the AASHTO soil classification system.

Classification of Sands and Gravels

In the USCS, gravels (given the symbol *G*) and sands (given the symbol *S*) are classified according to their grain size distribution. For the USCS, gravels may be given the classification symbol *GW* (well-graded gravel), *GP* (poorly graded gravel), *GM* (gravel with a large amount of silt), or *GC* (gravel with a large amount of clay). Likewise sands may be classified as being *SW*, *SP*, *SM* or *SC*. Sands and gravels with a small but non-negligible amount of fines (5–12%) may be given a dual classification such as *SW-SC*.

Atterberg Limits

Clays and Silts, often called 'fine-grained soils', are classified according to their Atterberg limits; the most commonly used Atterberg limits are the Liquid Limit (denoted by *LL* or w_l), Plastic Limit (denoted by *PL* or w_p), and Shrinkage Limit (denoted by *SL*).

The Liquid Limit is the water content at which the soil behavior transitions from a plastic solid to a liquid. The Plastic Limit is the water content at which the soil behavior transitions from that of a plastic solid to a brittle solid. The Shrinkage Limit corresponds to a water content below which the soil will not shrink as it dries. The consistency of fine grained soil varies in proportional to the water content in a soil.

As the transitions from one state to another are gradual, the tests have adopted arbitrary definitions to determine the boundaries of the states. The liquid limit is determined by measuring the water content for which a groove closes after 25 blows in a standard test. Alternatively, a fall cone test apparatus may be use to measure the liquid limit. The undrained shear strength of remolded soil at the liquid limit is approximately 2 kPa. The Plastic Limit is the water content below which it is not possible to roll by hand the soil into 3 mm diameter cylinders. The soil cracks or breaks up as it is rolled down to this diameter. Remolded soil at the plastic limit is quite stiff, having an undrained shear strength of the order of about 200 kPa.

The Plasticity Index of a particular soil specimen is defined as the difference between the Liquid Limit and the Plastic Limit of the specimen; it is an indicator of how much water the soil particles in the specimen can absorb, and correlates with many engineering properties like permeability, compressibility, shear strength and others. Generally, the clay having high plasticity have lower permeability and also they are also difficult to be compacted.

Classification of Silts and Clays

According to the Unified Soil Classification System (USCS), silts and clays are classified by plotting the values of their plasticity index and liquid limit on a plasticity chart. The A-Line on the chart separates clays (given the USCS symbol *C*) from silts (given the symbol *M*). LL=50% separates high plasticity soils (given the modifier symbol *H*) from low plasticity soils (given the modifier symbol *L*). A soil that plots above the A-line and has LL>50% would, for example, be classified as *CH*. Other possible classifications of silts and clays are *ML*, *CL* and *MH*. If the Atterberg limits plot in the"hatched" region on the graph near the origin, the soils are given the dual classification 'CL-ML'.

Indices Related to Soil Strength

Liquidity Index

The effects of the water content on the strength of saturated remolded soils can be quantified by the use of the *liquidity index, LI*:

$$LI = \frac{w - PL}{LL - PL}$$

when the LI is 1, remolded soil is at the liquid limit and it has an undrained shear strength of about 2 kPa. When the soil is at the plastic limit, the LI is 0 and the undrained shear strength is about 200 kPa.

Relative Density

The density of sands (cohesionless soils) is often characterized by the relative density, D_r

$$D_r = \frac{e_{max} - e}{e_{max} - e_{min}} 100\%$$

where: e_{max} is the "maximum void ratio" corresponding to a very loose state, e_{min} is the "minimum void ratio" corresponding to a very dense state and e is the *in situ* void ratio. Methods used to calculate relative density are defined in ASTM D4254-00(2006).

Thus if $D_r = 100\%$ the sand or gravel is very dense, and if $D_r = 0\%$ the soil is extremely loose and unstable.

Seepage: Steady State Flow of Water

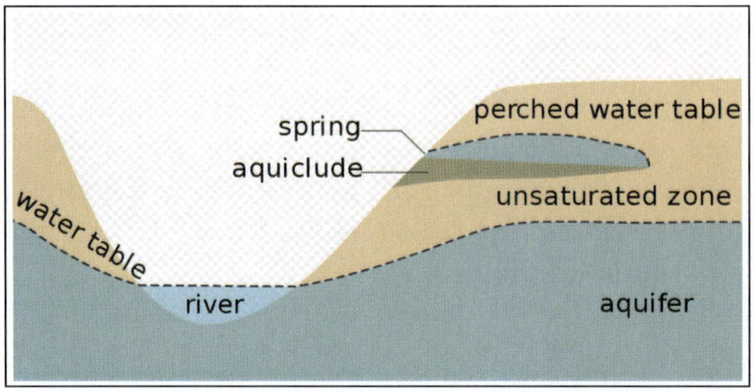

A cross section showing the water table varying with surface topography as well as a perched water table.

If fluid pressures in a soil deposit are uniformly increasing with depth according to $u = \rho_w g z_w$ then hydrostatic conditions will prevail and the fluids will not be flowing through the soil. z_w is the depth below the water table. However, if the water table is sloping or there is a perched water table as indicated in the accompanying sketch, then seepage will occur. For steady state seepage, the seepage velocities are not varying with time. If the water tables are changing levels with time, or if the soil is in the process of consolidation, then steady state conditions do not apply.

Darcy's Law

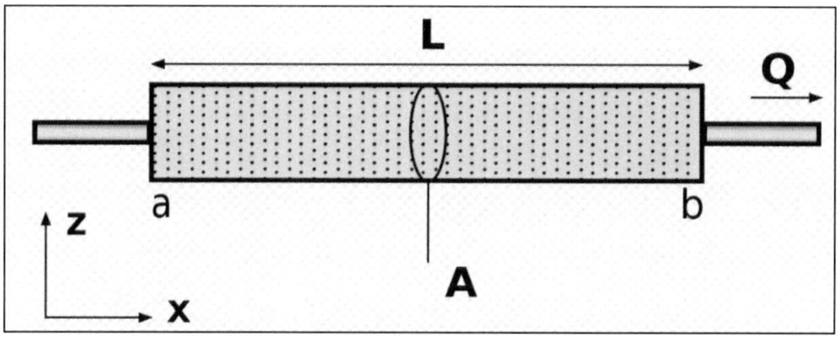

Diagram showing definitions and directions for Darcy's law.

Darcy's law states that the volume of flow of the pore fluid through a porous medium per unit time is proportional to the rate of change of excess fluid pressure with distance. The constant of proportionality includes the viscosity of the fluid and the intrinsic permeability of the soil. For the simple case of a horizontal tube filled with soil:

$$Q = \frac{-KA}{\mu}\frac{(u_b - u_a)}{L}.$$

The total discharge, Q (having units of volume per time, e.g., ft³/s or m³/s), is proportional to the intrinsic permeability, K, the cross sectional area, A, and rate of pore pressure change with distance, $\frac{u_b - u_a}{L}$, and inversely proportional to the dynamic viscosity of the fluid, μ. The negative sign is needed because fluids flow from high pressure to low pressure. So if the change in pressure is negative (in the x-direction) then the flow will be positive (in the x-direction). The above equation works well for a horizontal tube, but if the tube was inclined so that point b was a different elevation than point a, the equation would not work. The effect of elevation is accounted for by replacing the pore pressure by *excess pore pressure*, u_e defined as:

$$u_e = u - \rho_w g z$$

where z is the depth measured from an arbitrary elevation reference (datum). Replacing u by u_e we obtain a more general equation for flow:

$$Q = \frac{-KA}{\mu}\frac{(u_{e,b} - u_{e,a})}{L}$$

Dividing both sides of the equation by A, and expressing the rate of change of excess pore pressure as a derivative, we obtain a more general equation for the apparent velocity in the x-direction:

$$v_x = \frac{-K}{\mu}\frac{du_e}{dx}$$

where $v_x = Q / A$ has units of velocity and is called the *Darcy velocity* (or the specific discharge, filtration velocity, or superficial velocity). The *pore* or *interstitial velocity* v_{px} is the average velocity of fluid molecules in the pores; it is related to the Darcy velocity and the porosity n through the Dupuit-Forchheimer relationship:

$$v_{px} = \frac{v_x}{n}$$

(Some authors use the term *seepage velocity* to mean the Darcy velocity, while others use it to mean the pore velocity.)

Civil engineers predominantly work on problems that involve water and predominantly work on problems on earth (in earth's gravity). For this class of problems, civil engineers will often write Darcy's law in a much simpler form:

$$v = ki$$

where k is the hydraulic conductivity, defined as $k = \frac{K\rho_w g}{\mu_w}$, and i is the hydraulic gradient. The hydraulic gradient is the rate of change of total head with distance. The total head, h at a point is defined as the height (measured relative to the datum) to which water would rise in a piezometer at that point. The total head is related to the excess water pressure by:

$$u_e = \rho_w gh + Constant$$

and the *Constant* is zero if the datum for head measurement is chosen at the same elevation as the origin for the depth, z used to calculate u_e.

Typical values of hydraulic conductivity

Values of hydraulic conductivity, k, can vary by many orders of magnitude depending on the soil type. Clays may have hydraulic conductivity as small as about $10^{-12}\frac{m}{s}$, gravels may have hydraulic conductivity up to about $10^{-1}\frac{m}{s}$. Layering and heterogeneity and disturbance during the sampling and testing process make the accurate measurement of soil hydraulic conductivity a very difficult problem.

Flownets

Darcy's Law applies in one, two or three dimensions. In two or three dimensions, steady state seepage is described by Laplace's equation. Computer programs are available to solve this equation. But traditionally two-dimensional seepage problems were solved using a graphical procedure known as flownet. One set of lines in the flownet are in the direction of the water flow (flow lines), and the other set of lines are in the direction of constant total head (equipotential lines). Flownets may be used to estimate the quantity of seepage under dams and sheet piling.

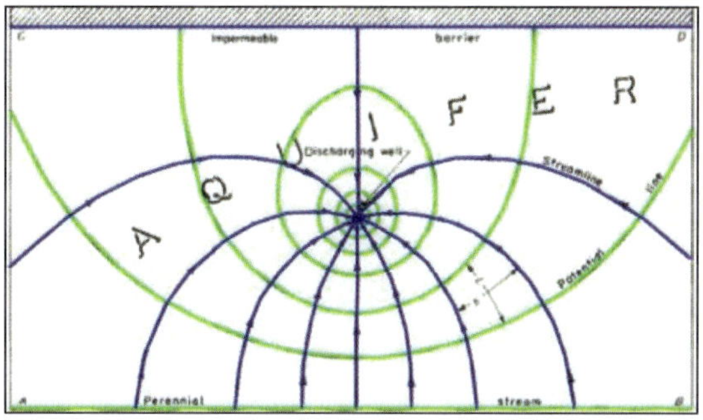

A plan flow net to estimate flow of water from a stream to a discharging well.

Seepage Forces and Erosion

When the seepage velocity is great enough, erosion can occur because of the frictional drag exerted on the soil particles. Vertically upwards seepage is a source of danger on the downstream side of sheet piling and beneath the toe of a dam or levee. Erosion of the soil, known as "soil piping", can lead to failure of the structure and to sinkhole formation. Seeping water removes soil, starting from the exit point of the seepage, and erosion advances upgradient. The term "sand boil" is used to describe the appearance of the discharging end of an active soil pipe.

Seepage Pressures

Seepage in an upward direction reduces the effective stress within the soil. When the water pressure at a point in the soil is equal to the total vertical stress at that point, the effective stress is zero and the soil has no frictional resistance to deformation. For a surface layer, the vertical effective stress becomes zero within the layer when the upward hydraulic gradient is equal to the critical gradient. At zero effective stress soil has very little strength and layers of relatively impermeable soil may heave up due to the underlying water pressures. The loss in strength due to upward seepage is a common contributor to levee failures. The condition of zero effective stress associated with upward seepage is also called liquefaction, quicksand, or a boiling condition. Quicksand was so named because the soil particles move around and appear to be 'alive' (the biblical meaning of 'quick' – as opposed to 'dead').

Effective Stress and Capillarity: Hydrostatic Conditions

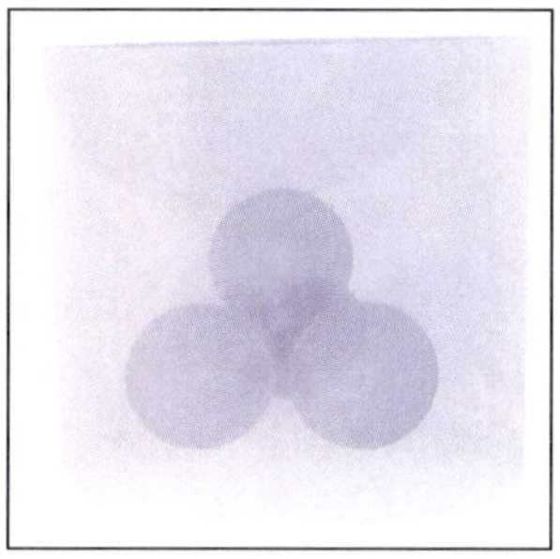

Spheres immersed in water, reducing effective stress.

To understand the mechanics of soils it is necessary to understand how normal stresses and shear stresses are shared by the different phases. Neither gas nor liquid provide significant resistance to shear stress. The shear resistance of soil is provided by friction and interlocking of the particles. The friction depends on the intergranular contact stresses between solid particles. The normal stresses, on the other hand, are shared by the fluid and the particles. Although the pore air is relatively compressible, and hence takes little normal stress in most geotechnical problems, liquid water is relatively incompressible and if the voids are saturated with water, the pore water must be squeezed out in order to pack the particles closer together.

The principle of effective stress, introduced by Karl Terzaghi, states that the effective stress σ› (i.e., the average intergranular stress between solid particles) may be calculated by a simple subtraction of the pore pressure from the total stress:

$$\sigma' = \sigma - u$$

where σ is the total stress and u is the pore pressure. It is not practical to measure σ› directly, so in practice the vertical effective stress is calculated from the pore pressure and vertical total stress. The distinction between the terms pressure and stress is also important. By definition, pressure at a point is equal in all directions but stresses at a point can be different in different directions. In soil mechanics, compressive stresses and pressures are considered to be positive and tensile stresses are considered to be negative, which is different from the solid mechanics sign convention for stress.

Total Stress

For level ground conditions, the total vertical stress at a point, σ_v, on average, is the weight of everything above that point per unit area. The vertical stress beneath a uniform surface layer with density ρ,, and thickness H is for example:

$$\sigma_v = \rho g H = \gamma H$$

where g is the acceleration due to gravity, and γ is the unit weight of the overlying layer. If there are multiple layers of soil or water above the point of interest, the vertical stress may be calculated by summing the product of the unit weight and thickness of all of the overlying layers. Total stress increases with increasing depth in proportion to the density of the overlying soil.

It is not possible to calculate the horizontal total stress in this way. Lateral earth pressures are addressed elsewhere.

Pore Water Pressure

Hydrostatic Conditions

If there is no pore water flow occurring in the soil, the pore water pressures will be hydrostatic. The water table is located at the depth where the water pressure is equal to the atmospheric pressure. For hydrostatic conditions, the water pressure increases linearly with depth below the water table:

$$u = \rho_w g z_w$$

where ρ_w is the density of water, and z_w is the depth below the water table.

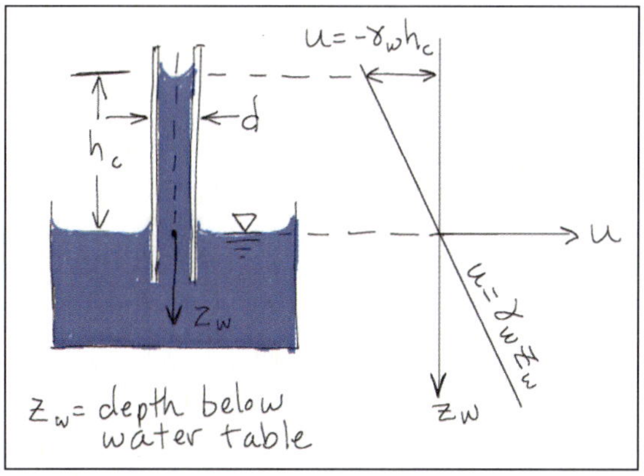

Water is drawn into a small tube by surface tension. Water pressure, u, is negative above and positive below the free water surface.

Capillary Action

Due to surface tension, water will rise up in a small capillary tube above a free surface of water. Likewise, water will rise up above the water table into the small pore spaces around the soil particles. In fact the soil may be completely saturated for some distance above the water table. Above the height of capillary saturation, the soil may be wet but the water content will decrease with elevation. If the water in the capillary zone is not moving, the water pressure obeys the equation of hydrostatic equilibrium, $u = \rho_w g z_w$, but note that z_w, is negative above the water table. Hence, hydrostatic water pressures are negative above the water table. The thickness of the zone of capillary saturation depends on the pore size, but typically, the heights vary between a centimeter or so for coarse sand to tens of meters for a silt or clay. In fact the pore space of soil is a uniform fractal e.g. a set of uniformly distributed D-dimensional fractals of average linear size L. For the clay soil it has been found that L=0.15 mm and D=2.7.

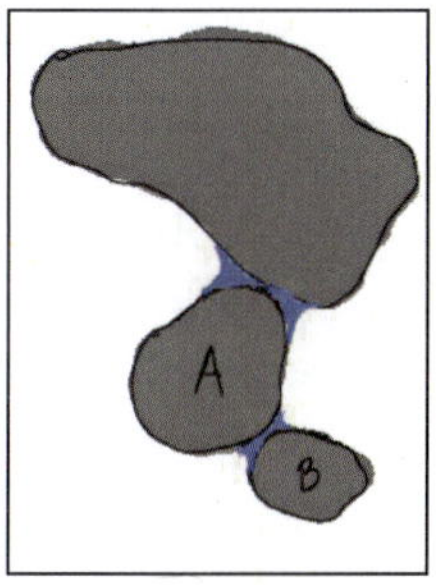

Water at particle contacts.

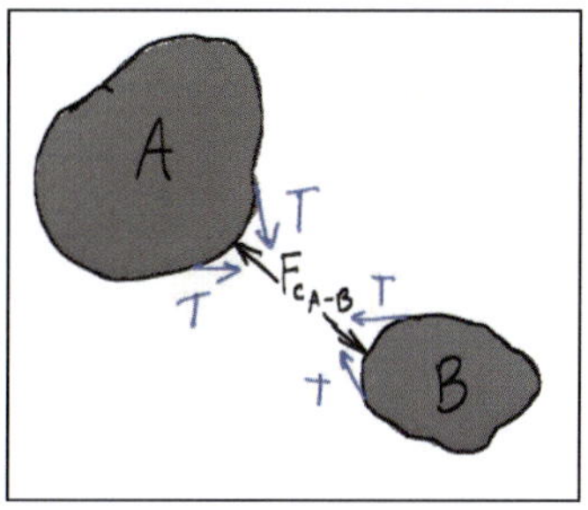

Intergranular contact force due to surface tension.

The surface tension of water explains why the water does not drain out of a wet sand castle or a moist ball of clay. Negative water pressures make the water stick to the particles and pull the particles to each other, friction at the particle contacts make a sand castle stable. But as soon as a wet sand castle is submerged below a free water surface, the negative pressures are lost and the castle collapses. Considering the effective stress equation, $\sigma' = \sigma - u$,, if the water pressure is negative, the effective stress may be positive, even on a free surface (a surface where the total normal stress is zero). The negative pore pressure pulls the particles together and causes compressive particle to particle contact forces. Negative pore pressures in clayey soil can be much more powerful than those in sand. Negative pore pressures explain why clay soils shrink when they dry and swell as they are wetted. The swelling and shrinkage can cause major distress, especially to light structures and roads.

Shrinkage caused by drying.

Consolidation: Transient Flow of Water

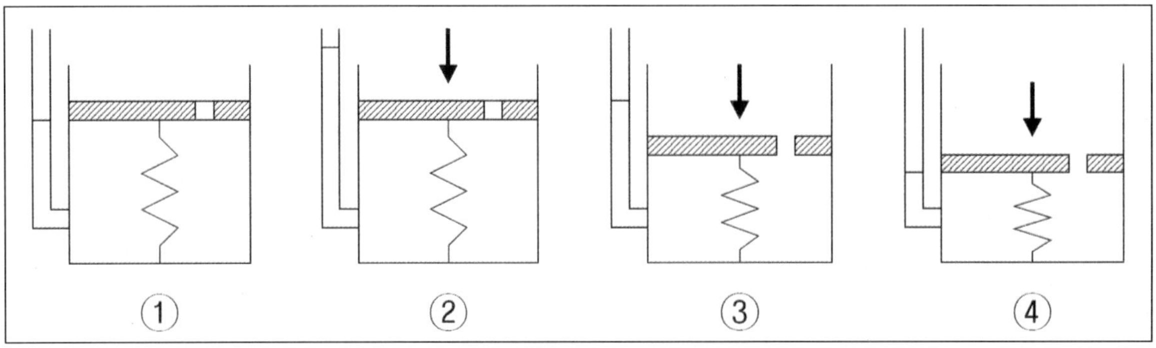

Consolidation analogy: The piston is supported by water underneath and a spring. When a load is applied to the piston, water pressure increases to support the load. As the water slowly leaks through the small hole, the load is transferred from the water pressure to the spring force.

Consolidation is a process by which soils decrease in volume. It occurs when stress is applied to a soil that causes the soil particles to pack together more tightly, therefore reducing volume. When this occurs in a soil that is saturated with water, water will be squeezed out of the soil. The time required to squeeze the water out of a thick deposit of clayey soil layer might be years. For a layer of sand, the water may be squeezed out in a matter of seconds. A building foundation or construction of a new embankment will cause the soil below to consolidate and this will cause settlement which in turn may cause distress to the building or embankment. Karl Terzaghi developed the theory of consolidation which enables prediction of the amount of settlement and the time required for the settlement to occur. Soils are tested with an oedometer test to determine their compression index and coefficient of consolidation.

When stress is removed from a consolidated soil, the soil will rebound, drawing water back into the pores and regaining some of the volume it had lost in the consolidation process. If the stress is re-applied, the soil will re-consolidate again along a recompression curve, defined by the recompression index. Soil that has been consolidated to a large pressure and has been subsequently unloaded is considered to be overconsolidated. The maximum past vertical effective stress is termed the preconsolidation stress. A soil which is currently experiencing the maximum past vertical effective stress is said to be normally consolidated. The overconsolidation ratio, (OCR) is the ratio of the maximum past vertical effective stress to the current vertical effective stress. The OCR is significant for two reasons: firstly, because the compressibility of normally consolidated soil is significantly larger than that for overconsolidated soil, and secondly, the shear behavior and dilatancy of clayey soil are related to the OCR through critical state soil mechanics; highly overconsolidated clayey soils are dilatant, while normally consolidated soils tend to be contractive.

Shear Behavior: Stiffness and Strength

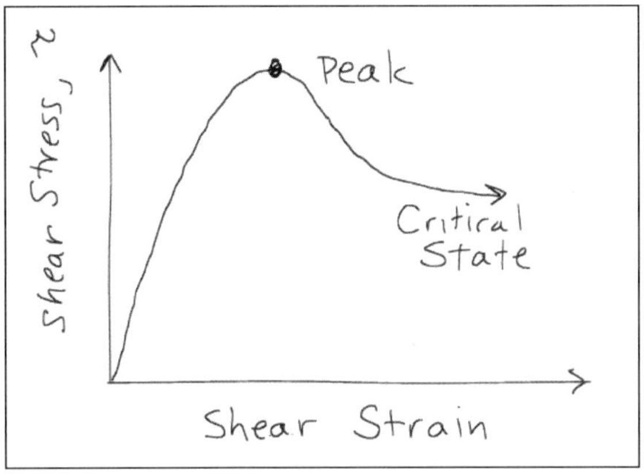

Typical stress strain curve for a drained dilatant soil.

The shear strength and stiffness of soil determines whether or not soil will be stable or how much it will deform. Knowledge of the strength is necessary to determine if a slope will be stable, if a building or bridge might settle too far into the ground, and the limiting pressures on a retaining wall. It is important to distinguish between failure of a soil element and the failure of a geotechnical structure (e.g., a building foundation, slope or retaining wall); some soil elements may reach their peak strength prior to failure of the structure. Different criteria can be used to define the "shear strength" and the "yield point" for a soil element from a stress–strain curve. One may define the peak shear strength as the peak of a stress–strain curve, or the shear strength at critical state as the value after large strains when the shear resistance levels off. If the stress–strain curve does not stabilize before the end of shear strength test, the "strength" is sometimes considered to be the shear resistance at 15–20% strain. The shear strength of soil depends on many factors including the effective stress and the void ratio.

The shear stiffness is important, for example, for evaluation of the magnitude of deformations of foundations and slopes prior to failure and because it is related to the shear wave velocity. The slope of the initial, nearly linear, portion of a plot of shear stress as a function of shear strain is called the shear modulus.

Friction, Interlocking and Dilation

Soil is an assemblage of particles that have little to no cementation while rock (such as sandstone) may consist of an assembly of particles that are strongly cemented together by chemical bonds. The shear strength of soil is primarily due to interparticle friction and therefore, the shear resistance on a plane is approximately proportional to the effective normal stress on that plane. The angle of internal friction is thus closely related to the maximum stable slope angle, often called the angle of repose.

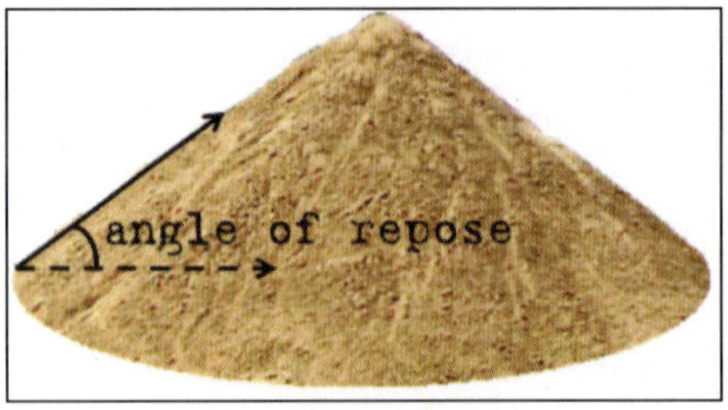

Angle of repose.

But in addition to friction, soil derives significant shear resistance from interlocking of grains. If the grains are densely packed, the grains tend to spread apart from each other as they are subject to shear strain. The expansion of the particle matrix due to shearing was called dilatancy by Osborne Reynolds. If one considers the energy required to shear an assembly of particles there is energy input by the shear force, T, moving a distance, x and there is also energy input by the normal force, N, as the sample expands a distance, y. Due to the extra energy required for the particles to dilate against the confining pressures, dilatant soils have a greater peak strength than contractive soils. Furthermore, as dilative soil grains dilate, they become looser (their void ratio increases), and their rate of dilation decreases until they reach a critical void ratio. Contractive soils become denser as they shear, and their rate of contraction decreases until they reach a critical void ratio.

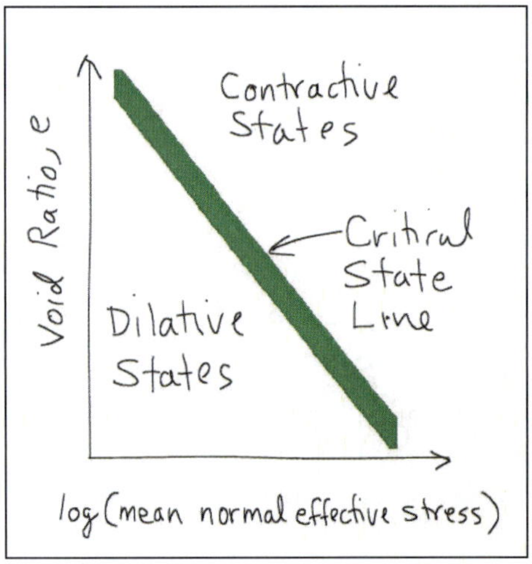

A critical state line separates the dilatant and contractive states for soil.

The tendency for a soil to dilate or contract depends primarily on the confining pressure and the void ratio of the soil. The rate of dilation is high if the confining pressure is small and the void ratio is small. The rate of contraction is high if the confining pressure is large and the void ratio is large. As a first approximation, the regions of contraction and dilation are separated by the critical state line.

Failure Criteria

After a soil reaches the critical state, it is no longer contracting or dilating and the shear stress on the failure plane τ_{crit} is determined by the effective normal stress on the failure plane σ'_n and critical state friction angle ϕ'_{crit}:

$$\tau_{crit} = \sigma'_n \tan\phi'_{crit}$$

The peak strength of the soil may be greater, however, due to the interlocking (dilatancy) contribution. This may be stated:

$$\tau_{peak} = \sigma'_n \tan\phi'_{peak}$$

where $\phi'_{peak} > \phi'_{crit}$. However, use of a friction angle greater than the critical state value for design requires care. The peak strength will not be mobilized everywhere at the same time in a practical problem such as a foundation, slope or retaining wall. The critical state friction angle is not nearly as variable as the peak friction angle and hence it can be relied upon with confidence.

Not recognizing the significance of dilatancy, Coulomb proposed that the shear strength of soil may be expressed as a combination of adhesion and friction components:

$$\tau_f = c' + \sigma'_f \tan\phi'$$

It is now known that the c' and ϕ' parameters in the last equation are not fundamental soil properties. In particular, c' and ϕ' are different depending on the magnitude of effective stress. According to Schofield, the longstanding use of c' in practice has led many engineers to wrongly believe that c' is a fundamental parameter. This assumption that c' and ϕ' are constant can lead to overestimation of peak strengths.

Structure, Fabric and Chemistry

In addition to the friction and interlocking (dilatancy) components of strength, the structure and fabric also play a significant role in the soil behavior. The structure and fabric include factors such as the spacing and arrangement of the solid particles or the amount and spatial distribution of pore water; in some cases cementitious material accumulates at particle-particle contacts. Mechanical behavior of soil is affected by the density of the particles and their structure or arrangement of the particles as well as the amount and spatial distribution of fluids present (e.g., water and air voids). Other factors include the electrical charge of the particles, chemistry of pore water, chemical bonds (i.e. cementation -particles connected through a solid substance such as recrystallized calcium carbonate).

Drained and Undrained Shear

The presence of nearly incompressible fluids such as water in the pore spaces affects the ability for the pores to dilate or contract.

If the pores are saturated with water, water must be sucked into the dilating pore spaces to fill the expanding pores (this phenomenon is visible at the beach when apparently dry spots form around feet that press into the wet sand).

Similarly, for contractive soil, water must be squeezed out of the pore spaces to allow contraction to take place.

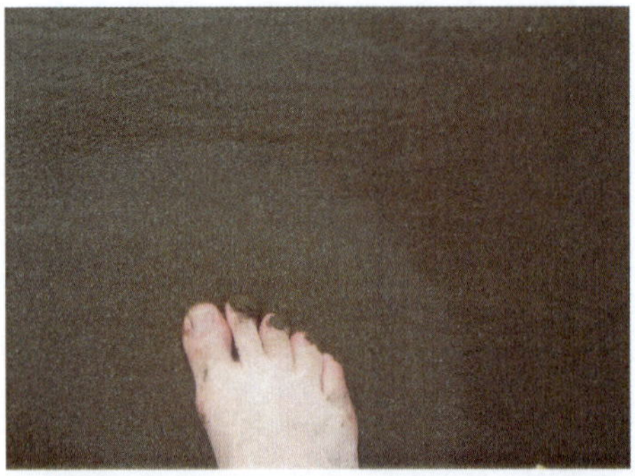

Foot pressure on sand causes it to dilate, drawing water from the surface into the pores.

Dilation of the voids causes negative water pressures that draw fluid into the pores, and contraction of the voids causes positive pore pressures to push the water out of the pores. If the rate of shearing is very large compared to the rate that water can be sucked into or squeezed out of the dilating or contracting pore spaces, then the shearing is called *undrained shear*, if the shearing is slow enough that the water pressures are negligible, the shearing is called *drained shear*. During undrained shear, the water pressure u changes depending on volume change tendencies. From the effective stress equation, the change in u directly effects the effective stress by the equation:

$$\sigma' = \sigma - u$$

and the strength is very sensitive to the effective stress. It follows then that the undrained shear strength of a soil may be smaller or larger than the drained shear strength depending upon whether the soil is contractive or dilative.

Shear Tests

Strength parameters can be measured in the laboratory using direct shear test, triaxial shear test, simple shear test, fall cone test and (hand) shear vane test; there are numerous other devices and variations on these devices used in practice today. Tests conducted to characterize the strength and stiffness of the soils in the ground include the Cone penetration test and the Standard penetration test.

Other Factors

The stress–strain relationship of soils, and therefore the shearing strength, is affected by:

- Soil composition (basic soil material): mineralogy, grain size and grain size distribution, shape of particles, pore fluid type and content, ions on grain and in pore fluid.
- State (initial): Define by the initial void ratio, effective normal stress and shear stress (stress history). State can be describe by terms such as: loose, dense, overconsolidated, normally consolidated, stiff, soft, contractive, dilative, etc.
- Structure: Refers to the arrangement of particles within the soil mass; the manner in which

the particles are packed or distributed. Features such as layers, joints, fissures, slickensides, voids, pockets, cementation, etc., are part of the structure. Structure of soils is described by terms such as: undisturbed, disturbed, remolded, compacted, cemented; flocculent, honey-combed, single-grained; flocculated, deflocculated; stratified, layered, laminated; isotropic and anisotropic.

- Loading conditions: Effective stress path -drained, undrained, and type of loading -magnitude, rate (static, dynamic), and time history (monotonic, cyclic).

Applications

Lateral Earth Pressure

Lateral earth stress theory is used to estimate the amount of stress soil can exert perpendicular to gravity. This is the stress exerted on retaining walls. A lateral earth stress coefficient, K, is defined as the ratio of lateral (horizontal) effective stress to vertical effective stress for cohesionless soils ($K=\sigma'_h/\sigma'_v$). There are three coefficients: at-rest, active, and passive. At-rest stress is the lateral stress in the ground before any disturbance takes place. The active stress state is reached when a wall moves away from the soil under the influence of lateral stress, and results from shear failure due to reduction of lateral stress. The passive stress state is reached when a wall is pushed into the soil far enough to cause shear failure within the mass due to increase of lateral stress. There are many theories for estimating lateral earth stress; some are empirically based, and some are analytically derived.

Bearing Capacity

The bearing capacity of soil is the average contact stress between a foundation and the soil which will cause shear failure in the soil. Allowable bearing stress is the bearing capacity divided by a factor of safety. Sometimes, on soft soil sites, large settlements may occur under loaded foundations without actual shear failure occurring; in such cases, the allowable bearing stress is determined with regard to the maximum allowable settlement. It is important during construction and design stage of a project to evaluate the subgrade strength. The California Bearing Ratio (CBR) test is commonly used to determine the suitability of a soil as a subgrade for design and construction. The field Plate Load Test is commonly used to predict the deformations and failure characteristics of the soil/subgrade and modulus of subgrade reaction (ks). The Modulus of subgrade reaction (ks) is used in foundation design, soil-structure interaction studies and design of highway pavements.

Slope Stability

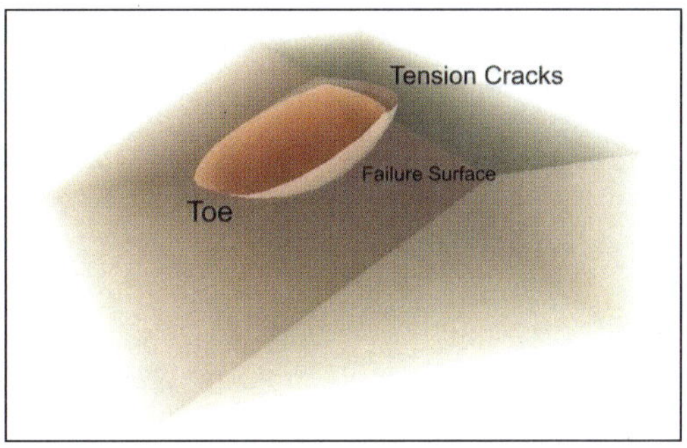

Simple slope slip section.

The field of slope stability encompasses the analysis of static and dynamic stability of slopes of earth and rock-fill dams, slopes of other types of embankments, excavated slopes, and natural slopes in soil and soft rock.

As seen to the right, earthen slopes can develop a cut-spherical weakness zone. The probability of this happening can be calculated in advance using a simple 2-D circular analysis package. A primary difficulty with analysis is locating the most-probable slip plane for any given situation. Many landslides have been analyzed only after the fact.

Soil Compaction

In geotechnical engineering, soil compaction is the process in which a stress applied to a soil causes densification as air is displaced from the pores between the soil grains. When stress is applied that causes densification due to water (or other liquid) being displaced from between the soil grains, then consolidation, not compaction, has occurred. Normally, compaction is the result of heavy machinery compressing the soil, but it can also occur due to the passage of (e.g.) animal feet.

In soil science and agronomy, soil compaction is usually a combination of both engineering compaction and consolidation, so may occur due to a lack of water in the soil, the applied stress being internal suction due to water evaporation as well as due to passage of animal feet. Affected soils become less able to absorb rainfall, thus increasing runoff and erosion. Plants have difficulty in compacted soil because the mineral grains are pressed together, leaving little space for air and water, which are essential for root growth. Burrowing animals also find it a hostile environment, because the denser soil is more difficult to penetrate. The ability of a soil to recover from this type of compaction depends on climate, mineralogy and fauna. Soils with high shrink-swell capacity, such as vertisols, recover quickly from compaction where moisture conditions are variable (dry spells shrink the soil, causing it to crack). But clays which do not crack as they dry cannot recover from compaction on their own unless they host ground-dwelling animals such as earthworms — the Cecil soil series is an example.

A crawler-backhoe is here equipped with a narrow sheepsfoot roller to compact the fill over newly placed sewer pipe, forming a stable support for a new road surface.

A compactor/roller fitted with a sheepsfoot drum, operated by U.S. Navy Seabees.

A Hamm vibrating roller with plain drum as used for compacting asphalt and granular soils.

Soil compaction is a vital part of the construction process. It is used for support of structural entities such as building foundations, roadways, walkways, and earth retaining structures to name a few. For a given soil type certain properties may deem it more or less desirable to perform adequately for a particular circumstance. In general, the preselected soil should have adequate strength, be relatively incompressible so that future settlement is not significant, be stable against volume change as water content or other factors vary, be durable and safe against deterioration, and possess proper permeability.

Wacker Neuson vibratory rammer BS 60-2i in action.

When an area is to be filled or backfilled the soil is placed in layers called lifts. The ability of the first fill layers to be properly compacted will depend on the condition of the natural material being covered. If unsuitable material is left in place and backfilled, it may compress over a long period under the weight of the earth fill, causing settlement cracks in the fill or in any structure supported by the fill. In order to determine if the natural soil will support the first fill layers, an area can be proofrolled. Proofrolling consists of utilizing a piece heavy construction equipment (typically, heavy compaction equipment or hauling equipment) to roll across the fill site and watching for deflections to be revealed. These areas will be indicated by the development of rutting, pumping, or ground weaving.

To ensure adequate soil compaction is achieved, project specifications will indicate the required soil density or degree of compaction that must be achieved. These specifications are generally recommended by a geotechnical engineer in a geotechnical engineering report.

The soil type - that is, grain-size distributions, shape of the soil grains, specific gravity of soil solids, and amount and type of clay minerals, present - has a great influence on the maximum dry unit weight and optimum moisture content. It also has a great influence on how the materials should be compacted in given situations. Compaction is accomplished by use of heavy equipment. In sands and gravels, the equipment usually vibrates, to cause re-orientation of the soil particles into a denser configuration. In silts and clays, a sheepsfoot roller is frequently used, to create small zones of intense shearing, which drives air out of the soil.

Determination of adequate compaction is done by determining the in-situ density of the soil and comparing it to the maximum density determined by a laboratory test. The most commonly used laboratory test is called the Proctor compaction test and there are two different methods in obtaining the maximum density. They are the standard Proctor and modified Proctor tests; the modified Proctor is more commonly used. For small dams, the standard Proctor may still be the reference.

While soil under structures and pavements needs to be compacted, it is important after construction to decompact areas to be landscaped so that vegetation can grow.

Compaction Methods

There are several means of achieving compaction of a material. Some are more appropriate for soil compaction than others, while some techniques are only suitable for particular soils or soils in particular conditions. Some are more suited to compaction of non-soil materials such as asphalt. Generally, those that can apply significant amounts of shear as well as compressive stress, are most effective.

The available techniques can be classified as:

- Static: A large stress is slowly applied to the soil and then released.
- Impact: The stress is applied by dropping a large mass onto the surface of the soil.
- Vibrating: A stress is applied repeatedly and rapidly via a mechanically driven plate or hammer. Often combined with rolling compaction.
- Gyrating: A static stress is applied and maintained in one direction while the soil is a subjected to a gyratory motion about the axis of static loading. Limited to laboratory applications.
- Rolling: A heavy cylinder is rolled over the surface of the soil. Commonly used on sports pitches. Roller-compactors are often fitted with vibratory devices to enhance their effectiveness.
- Kneading: Shear is applied by alternating movement in adjacent positions. An example, combined with rolling compaction, is the 'sheepsfoot' roller used in waste compaction at landfills.

The construction plant available to achieve compaction is extremely varied and is described elsewhere.

Test Methods in Laboratory

Soil compactors are used to perform test methods which cover laboratory compaction methods used to determine the relationship between molding water content and dry unit weight of soils. Soil placed as engineering fill is compacted to a dense state to obtain satisfactory engineering properties such as, shear strength, compressibility, or permeability. In addition, foundation soils are often compacted to improve their engineering properties. Laboratory compaction tests provide the basis for determining the percent compaction and molding water content needed to achieve the required engineering properties, and for controlling construction to assure that the required compaction and water contents are achieved. Test methods such as EN 13286-2, EN 13286-47, ASTM D698, ASTM D1557, AASHTO T99, AASHTO T180, AASHTO T193, BS 1377:4 provide soil compaction testing procedures.

Soil Gradation

Soil gradation is a classification of a coarse-grained soil that ranks the soil based on the different particle sizes contained in the soil. Soil gradation is an important aspect of soil mechanics and geotechnical engineering because it is an indicator of other engineering properties such as compressibility, shear strength, and hydraulic conductivity. In a design, the gradation of the in situ or on site soil often controls the design and ground water drainage of the site. A poorly graded soil will have better drainage than a well graded soil.

Soil is graded as either well graded or poorly graded. Soil gradation is determined by analyzing the results of a sieve analysis or a hydrometer analysis. The process for grading a soil is in accordance with either the Unified Soil Classification System or the AASHTO Soil Classification System. Gradation of a soil is determined by reading the grain size distribution curve produced from the results of laboratory tests on the soil. Gradation of a soil can also be determined by calculating the coefficient of uniformity, Cu, and the coefficient of curvature, Cc, of the soil and comparing the calculated values with published gradation limits.

Soil gradation is a classification of the particle size distribution of a soil. Coarse-grained soils, mainly gravels or sands, are graded as either well graded or poorly graded. Poorly graded soils are further divided into uniformly-graded or gap-graded soils. Fine-grained soils, mainly silts and clays, are classified according to their Atterberg limits.

Well Graded

A well graded soil is a soil that contains particles of a wide range of sizes and has a good representation of all sizes from the No. 4 to No. 200 sieves. A well graded gravel is classified as GW while a well graded sand is classified as SW.

Poorly Graded

A poorly graded soil is a soil that does not have a good representation of all sizes of particles from the No. 4 to No. 200 sieve. A poorly graded gravel is classified as GP while a poorly graded sand is classified as SP. Poorly graded soils are more susceptible to soil liquefaction than well graded soils.

A gap-graded soil is a soil that has an excess or deficiency of certain particle sizes or a soil that has at least one particle size missing. An example of a gap-graded soil is one in which sand of the No. 10 and No. 40 sizes are missing, and all the other sizes are present.

Process of Grading a Soil

The process of grading a soil is in accordance with either the Unified Soil Classification System or the AASHTO Soil Classification System. The steps in grading a soil are data collection, calculating coefficients of uniformity and curvature, and grading the soil based on the grading criteria given in the used soil classification system.

Data Collection

Soil gradation is determined by analyzing the results of a sieve analysis or a hydrometer analysis. In a sieve analysis, a coarse-grained soil sample is shaken through a series of woven-wire square-mesh sieves. Each sieve has successively smaller openings so particles larger than the size of each sieve are retained on the sieve. The percentage of each soil size is measured by weighing the amount retained on each sieve and comparing the weight to the total weight of the sample. The results of a sieve analysis are plotted as a grain size distribution curve, which is then analyzed to determine the soil gradation of the particular soil.

In a hydrometer analysis, a fine-grained soil sample is left to settle in a viscous fluid. This method is used based on Stoke's Law which relates terminal velocity of fall of a particle in a viscous fluid to the grain diameter and density of the grain in suspension. Grain diameter is calculated from a known distance and time of the fall of the particle. This is used to classify fine-grained soils.

Calculating the Coefficients of Uniformity and Curvature

Calculating the coefficients of uniformity and curvature requires grain diameters. The grain diameter can be found for each percent of the soil passing a particular sieve. This means that if 40% of the sample is retained on the No. 200 sieve then there is 60% passing the No. 200 sieve.

The coefficient of uniformity, C_u is a crude shape parameter and is calculated using the following equation:

$$C_u = \frac{D_{60}}{D_{10}}$$

where D_{60} is the grain diameter at 60% passing, and D_{10} is the grain diameter at 10% passing.

The coefficient of curvature, C_c is a shape parameter and is calculated using the following equation:

$$C_c = \frac{(D_{30})^2}{D_{10} \times D_{60}}$$

where D_{60} is the grain diameter at 60% passing, D_{30} is the grain diameter at 30% passing, and D_{10} is the grain diameter at 10% passing

Once the coefficient of uniformity and the coefficient of curvature have been calculated, they must be compared to published gradation criteria.

Criteria for Grading Soils

The following criteria are in accordance with the Unified Soil Classification System:

For a gravel to be classified as well graded, the following criteria must be met:

$$C_u > 4 \; \& \; 1 < C_c < 3$$

If both of these criteria are not met, the gravel is classified as poorly graded or GP. If both of these criteria are met, the gravel is classified as well graded or GW.

For a sand to be classified as well graded, the following criteria must be met:

$$C_u \geq 6 \; \& \; 1 < C_c < 3$$

If both of these criteria are not met, the sand is classified as poorly graded or SP. If both of these criteria are met, the sand is classified as well graded or SW.

Importance

Soil gradation is very important to geotechnical engineering. It is an indicator of other engineering properties such as compressibility, shear strength, and hydraulic conductivity.

In a design, the gradation of the in situ or on site soil often controls the design and ground water drainage of the site. A poorly graded soil will have better drainage than a well graded soil because there are more void spaces in a poorly graded soil.

When a fill material is being selected for a project such as a highway embankment or earthen dam, the soil gradation is considered. A well graded soil is able to be compacted more than a poorly graded soil. These types of projects may also have gradation requirements that must be met before the soil to be used is accepted.

When options for ground remediation techniques are being selected, the soil gradation is a controlling factor.

Soil Nailing

Soil nailing is a construction remedial measure to treat unstable natural soil slopes or as a construction technique that allows the safe over-steepening of new or existing soil slopes. The technique involves the insertion of relatively slender reinforcing elements into the slope – often general purpose reinforcing bars (rebar) although proprietary solid or hollow-system bars are also available. Solid bars are usually installed into pre-drilled holes and then grouted into place using a separate grout line, whereas hollow bars may be drilled and grouted simultaneously by the use of a sacrificial drill bit and by pumping grout down the hollow bar as drilling progresses. Kinetic methods of firing relatively short bars into soil slopes have also been developed. Bars installed using drilling techniques are usually fully grouted and installed at a slight downward inclination with bars installed at regularly spaced points across the slope face. A rigid facing (often pneumatically applied concrete, otherwise known as shotcrete) or isolated soil nail head plates may be used at the surface. Alternatively a flexible reinforcing mesh may be held against the soil face beneath the head plates. Rabbit proof wire mesh and environmental erosion control fabrics and may be used in conjunction with flexible mesh facing where environmental conditions dictate.

Soil nail components may also be used to stabilize retaining walls or existing fill slopes (embankments and levees); this is normally undertaken as a remedial measure.

Since its first application using modern techniques in Versailles in 1972, soil nailing is now a well-established technique around the world.

Preliminary Analysis

Four main points to be considered in determining if soil nailing would be an effective retention technique are as follows. First, the existing ground conditions should be examined. Next, the advantages and disadvantages for a soil nail wall should be assessed for the particular application being considered. Then other systems should be considered for the particular application. Finally, cost of the soil nail wall should be considered. Soil nail walls can be used for a variety of soil types and conditions. The most favorable conditions for soil nailing are as follows: The soil should be able to stand unsupported one to two meters high for a minimum of two days when cut vertical or nearly vertical. Also all soil nails within a cross section should be located above the groundwater table. If the soil nails are not located above the groundwater table, the groundwater should not negatively affect the face of the excavation, the bond between the ground and the soil nail itself. Based upon these favorable conditions for soil nailing stiff to hard fine-grained soils which include stiff to hard clays, clayey silts, silty clays, sandy clays, and sandy silts are preferred soils. Sand and gravels which are dense to very dense soils with some apparent cohesion also work well for soil nailing. Weathered rock is also acceptable as long as the rock is weathered evenly throughout (meaning no weakness planes). Finally, glacial soils work well for soil nailing.

A list of unfavorable or difficult soil conditions for soil nailing can include dry, poorly graded cohesion-less soils, soils with a high groundwater table, soils with cobbles and boulders, soft to very soft fine-grained soils, highly corrosive soils, weathered rock with unfavorable weakness planes, and loess. Other difficult conditions include prolonged exposure to freezing temperatures, a climate that has a repeated freeze-and-thaw cycle, and granular soils that are very loose.

Design

After a preliminary analysis of the site, initial designs of the soil nail wall can be begin. This begins with a selection of limit states and design approaches. The two most common limit states used in soil nail wall design is strength limit and service limit states. The strength limit state is the limit state that addresses potential failure mechanisms or collapse states of the soil nail wall system. The service limit state is the limit state that addresses loss of service function resulting from excessive wall deformation and is defined by restrictions in stress, deformation and facing crack width under regular service conditions. The two most common design approaches for soil nail walls are limit state design and service load design.

Initial design considerations include wall layout (wall height and length), soil nail vertical and horizontal spacing, soil nail pattern on wall face, soil nail inclination, soil nail length and distribution, soil nail material and relevant ground properties. With all these variables in the mind of the design engineer the next step is to use simplified charts to preliminarily evaluate nail length and maximum nail force. Nail length, diameter and spacing typically control external and internal stability of the wall. These parameters can be adjusted during design until all external and internal stability requirements are met. After the initial design is completed, final design progresses where the soil nail wall has to be tested for external and internal failure modes, seismic considerations and aesthetic qualities. Drainage, frost penetration and external loads such as wind and hydrostatic forces also have to be determined and included in the final examination of the design. Soil nail walls are not ideal in locations with highly plastic clay soils. Soils with high plasticity, a high liquid limit and low undrained shear strengths are at risk of long-term deformation (creep).

Construction

With the design complete, construction is the next step. Most soil nail wall construction follows a specific procedure. First a cut is excavated and temporary bracing is put in place if necessary. This is done with conventional earth moving equipment and hydraulic drills. Next, holes for the soil nails are drilled at predetermined locations as specified by the design engineer. The equipment used for this step is dependent on the stability of the material in which the soil nail wall is supporting. Rotary or rotary percussive methods using air flush or dry auger methods can be used with stable ground. For unstable ground, single tube and duplex rotary methods with air and water flush or hollow stem auger methods are used. With the holes drilled, the next step is to install and grout the nails into place. After all nails are inserted, a drainage system is put into place. Synthetic drainage mat is placed vertically between the nail heads, which are extended down to the base of the wall where they are most commonly connected to a footing drain. A layer of shotcrete is applied and bearing plates are installed before a final facing is put in place to complete the soil nail wall.

In terms of construction, soil nail walls have a decisive advantage over other alternatives. Soil nail walls require a smaller right-of-way than ground anchor walls and have less environmental impact. Installation of soil nail walls is relatively rapid and typically uses less materials and smaller construction equipment than ground anchor walls.

Cost Comparison

One great advantage of soil nail walls is their cost-effectiveness over other alternatives. When conventional soil nailing construction procedures are used, soil nail walls are much more economical than concrete gravity walls and similarly or more cost effective than ground anchor walls.

Inspection and Performance Monitoring

Inspection activities play a vital role in the production of high-quality soil nail walls because conformance to project plans and specifications should result in a soil nail wall that will perform its intended duty for its designed duration. Inspections usually involve evaluation of the following: conformance of system components to material specification, conformance of construction methods to execution specifications, conformance to short-term performance specifications, and long-term monitoring. Short-term performance specifications are checked with loads tests, which use hydraulic jacks and pumps to perform several load applications. Three common load tests for short-term performance are verification or ultimate load tests, proof tests and creep tests. Verification or ultimate load tests are conducted to verify the compliance of the soil nails with pullout capacity and strengths resulting from the contractor's installation method. Proof tests are intended to verify that the contractor's construction procedure has been consistent and that the nails have not been drilled and grouted in a soil zone not tested in the verification stage. Creep tests are performed to ensure that the nail design loads can be safely carried throughout the structure's service life.

Long-term performance monitoring is used to collect data to ensure adequate performance and refine future design practices. Parameters to be measured include vertical and horizontal movement of the wall face, local movements or deterioration of facing elements, drainage to the ground, loads, load distribution and load changes in the nails, temperature and rainfall. These parameters are measured using several specific tools including inclinometers, load cells and strain gauges.

Erosion Control

Erosion control is the practice of preventing or controlling wind or water erosion in agriculture, land development, coastal areas, river banks and construction. Effective erosion controls handle surface runoff and are important techniques in preventing water pollution, soil loss, wildlife habitat loss and human property loss.

Usage

Erosion controls are used in natural areas, agricultural settings or urban environments. In urban areas erosion controls are often part of stormwater runoff management programs required by local governments. The controls often involve the creation of a physical barrier, such as vegetation or rock, to absorb some of the energy of the wind or water that is causing the erosion. They also involve building and maintaining storm drains. On construction sites they are often implemented in conjunction with sediment controls such as sediment basins and silt fences.

Bank erosion is a natural process: without it, rivers would not meander and change course. However, land management patterns that change the hydrograph and/or vegetation cover can act to increase or decrease channel migration rates. In many places, whether or not the banks are unstable due to human activities, people try to keep a river in a single place. This can be done for environmental reclamation or to prevent a river from changing course into land that is being used by people. One way that this is done is by placing riprap or gabions along the bank.

Examples of erosion control methods include:

- Buffer strip

- Cellular confinement systems
- Crop rotation
- Conservation tillage
- Contour bunding
- Contour plowing
- Cover crops
- Fiber rolls
- Gabions
- Hydroseeding
- Level spreaders
- Mulching
- Perennial crops
- Plasticulture
- Polyacrylamide (as a coagulant)
- Reforestation
- Riparian buffer
- Riprap
- Strip farming
- Sand fence
- Vegetated waterway (bioswale)
- Terracing
- Wattle (construction)
- Windbreaks

Mathematical Modeling

Since the 1920s and 1930s scientists have been creating mathematical models for understanding the mechanisms of soil erosion and resulting sediment surface runoff, including an early paper by Albert Einstein applying Baer's law. These models have addressed both gully and sheet erosion. Earliest models were a simple set of linked equations which could be employed by manual calculation. By the 1970s the models had expanded to complex computer models addressing nonpoint source pollution with thousands of lines of computer code. The more complex models were able to address nuances in micrometeorology, soil particle size distributions and micro-terrain variation.

Berm

A berm is a level space, shelf, or raised barrier (usually made of compacted soil) separating two areas. It can serve as a fortification line, a border/separation barrier, in industrial settings, or in many other applications. A berm is also used on mountain bike and BMX courses to assist the rider around a corner at speed; this type of berm is often steep and curved, unlike most "construction berms". The word berm originates in the Middle Dutch and came into usage in English via French.

Military Use

In medieval military engineering, a berm (or berme) was a level space between a parapet or defensive wall and an adjacent steep-walled ditch or moat. It was intended to reduce soil pressure on the walls of the excavated part to prevent its collapse. It also meant that debris dislodged from fortifications would not fall into (and fill) a ditch or moat.

In the trench warfare of World War I, the name was applied to a similar feature at the lip of a trench, which served mainly as an elbow-rest for riflemen.

Modern Usage

In modern military engineering, a berm is the earthen or sod wall or parapet itself, especially a low earthen wall adjacent to a ditch. The digging of the ditch (often by a bulldozer or military engineering vehicle) can provide the soil from which the berm is constructed. Walls constructed in this manner are an effective obstacle to vehicles, including most armoured fighting vehicles, but are easily crossed by infantry. Because of the ease of construction, such walls can be made hundreds or thousands of kilometres long. A prominent example of such a berm is the 2,700 km (1,700 mi) Moroccan Western Sahara Wall.

Erosion Control

Berms are also used to control erosion and sedimentation by reducing the rate of surface runoff. The berms either reduce the velocity of the water, or direct water to areas that are not susceptible to erosion, thereby reducing the adverse effects of running water on exposed topsoil. Following the 2010 *Deepwater Horizon* oil spill in the Gulf of Mexico, the construction of berms designed to prevent oil from reaching the fragile Louisiana wetlands (which would result in massive erosion) was proposed early on, and was officially approved by the federal government in mid-June, 2010, after numerous failures to stop and contain the oil leak with more advanced technologies.

Geography

In coastal geography, a berm is a bank of sand or gravel ridge parallel to the shoreline and a few tens of centimetres high, created by wave action throwing material beyond the average level of the sea.

House Construction

Earth is piled up against exterior walls and packed, sloping down away from the house. The roof may or may not be fully earth covered, and windows/openings may occur on one or more sides of the shelter. Due to the building being above ground, fewer moisture problems are associated with earth berming in comparison to underground/fully recessed construction.

Other Applications

A typical berm application.

For general applications, a berm is a physical, stationary barrier of some kind. For example, in modern highway construction, a berm is a noise barrier constructed of earth, often landscaped, running along a highway to protect adjacent land users from noise pollution. The shoulder of a road is also called a berm and in New Zealand the word sometimes describes a publicly owned grassed nature strip sometimes planted with trees alongside urban roads (usually called a verge). In snowboard cross, a berm is a wall of snow built up in a corner. In mountain biking, a berm is a banked turn formed by soil, commonly dug from the track, being deposited on the outer rim of the turn. In coastal systems, a berm is a raised ridge of pebbles or sand found at high tide or storm tide marks on a beach. In snow removal, a berm or windrow refers to the linear accumulation of snow cast aside by a plow. In open-pit mining, a berm refers to dirt and rock piled alongside a haulage road or along the edge of a dump point. Intended as a safety measure, they are commonly required by government organizations to be at least half as tall as the wheels of the largest mining machine on-site.

Physical security systems employ berms to exclude hostile vehicles and slow attackers on foot (similar to the military application without the trench). Security berms are common around military and nuclear facilities. An example is the berm proposed for Vermont Yankee nuclear power plant in Vermont. At Baylor Ballpark, a baseball stadium on the campus of Baylor University, a berm is constructed down the right field line. The berm replaces bleachers, and general admission tickets are sold for fans who wish to sit on the grass or watch the game from the top of the hill.

Berms are also used as a method of environmental spill containment and liquid spill control. Bunding is the construction of a secondary impermeable barrier around and beneath storage or processing plant, sufficient to contain the plant's volume after a spill. This is often achieved on large sites by surrounding the plant with a berm. The US Environmental Protection Agency (EPA) requires that oils and fuels stored over certain volume levels be placed in secondary spill containment. Berms for spill containment are typically manufactured from polyvinyl chloride (PVC) or geomembrane fabric that provide a barrier to keep spills from reaching the ground or navigable waterways. Most berms have sidewalls to keep liquids contained for future capture and safe disposal.

Borrow Pit

An excavator working in a borrow pit.

In construction and civil engineering, a borrow pit, also known as a sand box, is an area where material (usually soil, gravel or sand) has been dug for use at another location. Borrow pits can be found close to many major construction projects. For example, soil might be excavated to fill an embankment for a highway, clay might be excavated for use in brick-making, gravel to be used for making concrete, etc.

In some cases, the borrow pits may become filled with ground water, forming recreational areas or sustainable wildlife habitats (one such example is the Merton Borrow Pit, near Oxford in central England, excavated to provide materials for the nearby M40 motorway). In other cases, borrow pits may be used for landfill and waste disposal.

A regional variation of this is termed "barrow pit" in the western United States (especially the Rocky Mountains). The localism describes the ditch along a roadway. These ditches were created to provide the fill to level and crown the roadway and subsequently provided drainage for the road.

An excavation lake (also a flooded gravel pit) is an artificial lake, which usually has its origins in the excavation of gravel or sand for construction materials or in some other kind of surface mining. In many cases, the excavation holes are landscaped according to the land restoration required by law. Because the excavation reached a point below the water table, lakes form naturally. Less frequently excavation lakes are intentionally created, especially as recreation areas.

Panorama of a "young" excavation lake in the Fürstenfeldbruck district of Germany.

In Germany the lakes are almost always used for fishing, since a fishery is created by law with every surface water. At some excavation lakes, beaches are added for swimming or other water sports, in particular boating, water skiing or windsurfing. To support these uses, large parking lots, changing areas, and eating areas are also set up. In some cases, the excavation lake serves as a nature reserve, as in the case of the lakes in the Attenborough Nature Reserve.

Caisson

In geotechnical engineering, a caisson is a watertight retaining structure used, for example, to work on the foundations of a bridge pier, for the construction of a concrete dam, or for the repair of ships. Caissons are constructed in such a way that the water can be pumped out, keeping the work environment dry. When piers are being built using an open caisson, and it is not practical to reach suitable soil, friction pilings may be driven to form a suitable sub-foundation. These piles are connected by a foundation pad upon which the column pier is erected.

Function

Shallow caissons may be open to the air, whereas pneumatic caissons, which penetrate soft mud, are sealed at the top and filled with compressed air to keep water and mud out at depth. An airlock allows access to the chamber. Workers move mud and rock debris (called *muck*) from the edge of the workspace to a water-filled pit, connected by a tube (called the *muck tube*) to the surface. A crane at the surface removes the soil with a clamshell bucket. The water pressure in the tube balances the air pressure, with excess air escaping up the muck tube. The pressurized air flow must be constant to ensure regular air changes for the workers and prevent excessive inflow of mud or water at the base of the caisson.

Installation

To install a caisson in place, it is brought down through soft mud until a suitable foundation material is encountered. While bedrock is preferred, a stable, hard mud is sometimes used when bedrock is too deep.

Dangers Encountered in Pressurized Caissons

Construction workers who leave the pressurized environment of the caisson must decompress at a rate that allows symptom-free release of inert gases dissolved in the body tissues if they are to avoid decompression sickness, a condition first identified in caisson workers, and originally named "caisson disease" in recognition of the occupational hazard. Construction of the Brooklyn Bridge, which was built with the help of pressurised caissons, resulted in numerous workers being either killed or permanently injured by caisson disease during its construction. Barotrauma of the ears, sinus cavities and lungs and dysbaric osteonecrosis are other risks.

Other Uses

Caissons have also been used in the installation of hydraulic elevators where a single-stage ram is installed below the ground level.

Caissons, codenamed *Phoenix*, were an integral part of the Mulberry harbours used during the World War II Allied invasion of Normandy.

Types

The four main types of caisson are box caisson, open caisson, pneumatic caisson and monolith caisson.

Box

A *box caisson* is a prefabricated concrete box (with sides and a bottom); it is set down on prepared bases. Once in place, it is filled with concrete to become part of the permanent works, such as the foundation for a bridge pier. Hollow concrete structures are usually less dense than water so a box caisson must be ballasted or anchored to keep it from floating until it can be filled with concrete. Sometimes elaborate anchoring systems may be required, such as in tidal zones. Adjustable anchoring systems combined with a GPS survey enable engineers to position a box caisson with pinpoint accuracy.

Open

An *open caisson* is similar to a box caisson, except that it does not have a bottom face. It is suitable for use in soft clays (e.g. in some river-beds), but not for where there may be large obstructions in the ground. An open caisson that is used in soft grounds or high water tables, where open trench excavations are impractical, can also be used to install deep manholes, pump stations and reception/launch pits for micro tunnelling, pipe jacking and other operations.

A caisson is sunk by self-weight, concrete or water ballast placed on top, or by hydraulic jacks. The leading edge (or *cutting shoe*) of the caisson is sloped out at a sharp angle to aid sinking in a vertical manner; it is usually made of steel. The shoe is generally wider than the caisson to reduce friction, and the leading edge may be supplied with pressurised bentonite slurry, which swells in water, stabilizing settlement by filling depressions and voids. An open caisson may fill with water during sinking. The material is excavated by clamshell excavator bucket on crane.

The formation level subsoil may still not be suitable for excavation or bearing capacity. The water in the caisson (due to a high water table) balances the upthrust forces of the soft soils underneath. If dewatered, the base may "pipe" or "boil", causing the caisson to sink. To combat this problem, piles may be driven from the surface to act as:

- Load-bearing walls, in that they transmit loads to deeper soils.
- Anchors, in that they resist flotation because of the friction at the interface between their surfaces and the surrounding earth into which they have been driven.

H-beam sections (typical column sections, due to resistance to bending in all axis) may be driven at angles "raked" to rock or other firmer soils; the H-beams are left extended above the base. A reinforced concrete plug may be placed under the water, a process known as *Tremie concrete placement*. When the caisson is dewatered, this plug acts as a pile cap, resisting the upward forces of the subsoil.

Pneumatic

A pneumatic (compressed-air) caisson has the advantage of providing dry working conditions, which is better for placing concrete. It is also well suited for foundations for which other methods might cause settlement of adjacent structures.

Monolithic

A monolithic caisson (or just simply a monolith) is larger than the other types of caisson, but similar to open caissons. Such caissons are often found in quay walls, where resistance to impact from ships is required.

Boat Lift Caissons

The word *caisson* is also used as a synonym for the moving trough part of caisson locks, canal lifts and inclines in which boats and ships rest while being lifted from one canal elevation to another; the water is retained on the inside of the caisson, or excluded from the caisson, according to the respective operating principle.

Structural Caissons

Caisson is also sometimes used as a colloquial term for a reinforced concrete structure formed by pouring into a hollow cylindrical form, typically by placing a caisson form below grade in an open excavation and pouring once backfill is complete, or by drilling at grade, although this can be problematic with deep caissons, as unsupported excavations can collapse before the caisson form can be inserted. In this manner, the earth placed around the empty caisson form provides stability and strength, allowing concrete to be poured with fewer complications and with less risk of a form blowout. While, technically, only the form itself is actually a caisson, it is not uncommon for any below-grade cast concrete pillar to be referred to as, simply, a *caisson*.

Ventilation Filtration Systems

The word *caisson* is also used as a name for an airtight housing for ventilation filters in facilities that handle hazardous materials. The housing usually has an upstream compartment for a pre-filter element and a downstream compartment for a high-efficiency filter element. It may have multiple sets of compartments. The housing has gasketed access doors to allow for the change out of the filter elements. The housing is usually equipped with connection points used to test the efficiency of the filters and monitor changes in the differential pressure across the filter media.

Cellular Confinement Systems

Cellular confinement systems (CCS)—also known as geocells—are widely used in construction for erosion control, soil stabilization on flat ground and steep slopes, channel protection, and structural reinforcement for load support and earth retention. Typical cellular confinement systems are geosynthetics made with ultrasonically welded high-density polyethylene (HDPE) strips or novel polymeric alloy (NPA)—and expanded on-site to form a honeycomb-like structure—and filled with sand, soil, rock, gravel or concrete.

Research and development of cellular confinement systems (CCS) began with the U.S. Army Corps of Engineers in 1975 to devise a method for building tactical roads over soft ground. Engineers found that sand-confinement systems performed better than conventional crushed stone sections and they could provide an expedient construction technique for access roads over soft ground, without being adversely affected by wet weather conditions. The US Army Corps of Engineers in Vicksburg, Mississippi (1981) experimented with a number of confining systems, from plastic pipe mats, to slotted aluminum sheets to prefabricated polymeric systems called sand grids and then, cellular confinement systems. Today cellular confinement systems are typically made from strips 50–200 mm wide, ultrasonically welded at intervals along their width. The CCS is folded and shipped to the job site in a collapsed configuration.

Efforts for civilian commercialization of the cellular confinement system by the Presto Products Company, led to the Geoweb. This cellular confinement system was made from high density polyethylene (HDPE), relatively strong, lightweight and suitable for geosynthetic extruding manufacturing. The cellular confinement system was used for load support, slope erosion control and channel lining and earth retention applications in the United States and Canada in the early 1980s.

Recent Innovations in Cellular Confinement Technology

The strength and stiffness of pavement layers determines the performance of highway pavements while aggregate use impacts the cost of duration of installation; therefore alternatives are needed to improve pavement quality using new materials with less aggregate usage. Geocells are recognized as a suitable geosynthetic reinforcement of granular soils to support static and moving wheel loads on roadways, railways and similar applications. But stiffness of the geocells was identified as a key influencing factor for geocell reinforcement, and hence the rigidity of the entire pavement structure.

Application vs. long-term Performance

CCS have been successfully installed in thousands of projects worldwide. However, it is incumbent to differentiate between low load applications, such as slope and channel applications, and new heavy-duty applications, such as in the base layer of asphalt pavement structures of heavily trafficked motorways and highways. While all polymeric materials used in CCS creep over time and under loading, the question is; what is the rate of degradation, under which conditions, how will this impact performance, and when will it fail?

The lifespan of CCS in slope protection applications, for example, is less critical as vegetative growth and root interlock stabilize the soil. This in effect compensates for any long-term loss of confinement in the CCS. Similarly, load support applications for low volume roads that are not subject to heavy loading usually have a short design life; therefore minor loss of performance is tolerable. However, in critical applications such as reinforcement of the structural layer of asphalt highway pavements, long-term dimensional stability is critical. The required design life for such roads under heavy traffic loads is typically 20–25 years, requiring verifiable long-term durability.

Development of standards for CCS

There were few test standards for geocells and fewer for their use in design. Test standards for

CCS were developed more than 40 years ago, while other test methods evolved from 2D planar geosynthetics. These do not reflect the composite behavior of 3D geometry of CCS, nor do they test long-term parameters such as: dynamic elastic stiffness, permanent plastic deformation and oxidation resistance. However, ISO/ASTM procedures have been developed for testing polymers in the space and automobile industries, as well as for other geosynthetic products. These new standards for CCS were proposed and under discussion by leading experts in geosynthetics in ASTM technical committee D-35. The stated goal is to set new industry standards that more accurately reflect 3D cellular confinement system geometry and material performance in the field rather than lab tests of individual strips and virgin materials that are typically used today.

A recent development in standards for the use of reinforcement geosynthetics in roadways was recently published by in the Netherlands. This standard covers geocell (as well as geogrid) applications, support mechanisms, and design principles. It also emphasizes the importance of the geocell material attributes (stiffness and creep resistance) and how they influence long-term reinforcement factors. Additional guidelines for the use of geocells in roadway applications are currently under development by the ISO and ASTM organizations, but have not yet been published.

How it Works?

A Cellular Confinement System when infilled with compacted soil creates a new composite entity that possesses enhanced mechanical and geotechnical properties. When the soil contained within a CCS is subjected to pressure, as in the case of a load support application, it causes lateral stresses on perimeter cell walls. The 3D zone of confinement reduces the lateral movement of soil particles while vertical loading on the contained infill results in high lateral stress and resistance on the cell-soil interface. These increase the shear strength of the confined soil, which:

- Creates a stiff mattress or slab to distribute the load over a wider area.
- Reduces punching of soft soil.
- Increases shear resistance and bearing capacity.
- Decreases deformation.

Confinement from adjacent cells provides additional resistance against the loaded cell through passive resistance, while lateral expansion of the infill is restricted by high hoop strength. Compaction is maintained by the confinement, resulting in long-term reinforcement.

On site, the geocell sections are fastened together and placed directly on the subsoil's surface or on a geotextile filter placed on the subgrade surface and propped open in an accordion-like fashion with an external stretcher assembly. The sections expand to an area of several tens of meters and consist of hundreds of individual cells, depending on the section and cell size. They are then filled with various infill materials, such as soil, sand, aggregate or recycled materials and then compacted using vibratory compactors. Surface layers many be of asphalt or unbound gravel materials.

Applications

Roadway Load Support

Cellular Confinement Systems (CCS) have been used to improve the performance of both paved and unpaved roads by reinforcing the soil in the subgrade-base interface or within the base course. The effective load distribution of CCS creates a strong, stiff cellular mattress. This 3D mattress reduces vertical differential settlement into soft subgrades, improves shear strength, and enhances load-bearing capacity, while reducing the amount of aggregate material required to extend the service life of roads. As a composite system, cellular confinement strengthens the aggregate infill, thereby simultaneously enabling the use of poorly graded inferior material (e.g. local native soils, quarry waste or recycled materials) for infill as well as reducing the structural support layer thickness. Typical load support applications include reinforcement of base and subbase layers in flexible pavements, including: asphalt pavements; unpaved access, service and haul roads; military roads, railway substructure and ballast confinement; working platforms in intermodal ports; airport runways and aprons, permeable pavements; pipeline road support; green parking facilities and emergency access areas.

Steep Soil Slope and Channel protection

The three-dimensional lateral confinement of CCS along with anchoring techniques ensures the long-term stability of slopes using vegetated topsoil, aggregate or concrete surfacing (if exposed to severe mechanical and hydraulic pressures). The enhanced drainage, frictional forces and cell-soil-plant interaction of CCS prevents downslope movement and limits the impact of raindrops, channeling and hydraulic shear stresses. The perforations in the 3D cells allow the passage of water, nutrients and soil organisms. This encourages plant growth and root interlock, which further stabilizes the slope and soil mass, and facilitates landscape rehabilitation. Typical applications include: construction cut and fill slopes and stabilization; road and rail embankments; pipeline stabilization and storage facility berms; quarry and mine site restoration; channel and coastline structures. They can be built as an underlying mass or as a facing.

Earth Retention

CCS provide steep vertical mechanically stabilized earth structures (either gravity or reinforced walls) for steep faces, walls and irregular topography. Construction of CCS earth retention is simiplified as each layer is structurally sound thereby providing access for equipment and workers, while eliminating the need for concrete formwork and curing. Local soil can be used for infill when suitable and granular, while the outer faces enable a green or tan fascia of the horizontal terraces/rows utilizing topsoil. Walls also can be used for lining channels and in cases of high flow, it is required that the outer cells contain concrete or cementious slurry infill. CCS have been used to reinforce soft or uneven soil foundations for large area footings, for retaining wall strip footings, for load sharing of covers over pipelines and other geotechnical applications.

Reservoirs and Landfills

CCS provides membrane liner protection, while creating stable soil, berms and slopes, for non-slip protection and durable impoundment of liquids and waste. Infill treatment depends on the contained materials: concrete for ponds and reservoirs; gravel for landfill drainage and leachates, vegetated infill for landscape rehabilitation. Concrete work is efficient and controlled as CCS functions as ready-made forms; CCS with concrete forms a flexible slab that accommodates minor subgrade movement and prevents cracking. In medium and low flow-velocities, CCS with geomembranes and gravel cover can be used to create impermeable channels, thereby eliminating the need for concrete.

Sustainable Construction

CCS is a green solution that makes civil infrastructure projects more sustainable. In load support applications, by reducing the amount and type of infill needed to reinforce soil, the usage of haul and earthmoving equipment is reduced. This in turn decreases fuel use, pollution and the carbon footprint, and at the same time minimizes on-site disruption from dust, erosion and runoff. When used for slope applications, perforated CCS provides excellent soil protection, water drainage and growth stratum for plants. The long-term design life of advanced CCS technology means that maintenance and the associated environmental costs are significantly reduced, as are long-term economic costs.

Core Recovery Parameters

Core recovery parameters describe the quality of core recovered from a borehole.

Total Core Recovery

Total core recovery (TCR) is the borehole core recovery percentage.

TCR is defined as the quotient:

$$TCR = \left(\frac{l_{\text{sum of pieces}}}{l_{\text{tot core run}}} \right) \times 100\%$$

$l_{\text{sum of pieces}}$ = Sum of length of core pieces.

$l_{\text{tot core run}}$ = Total length of core run.

Solid Core Recovery

Solid core recovery (SCR) is the borehole core recovery percentage of solid, cylindrical, pieces of rock core.

SCR is defined as the quotient:

$$SCR = \left(\frac{l_{\text{sum of solid core pieces}}}{l_{\text{tot core run}}} \right) \times 100\%$$

$l_{\text{sum of solid core pieces}}$ = Sum of length of solid, cylindrical, core pieces.

$l_{\text{tot core run}}$ = Total length of core run.

Rock Quality Designation

Rock-quality designation (RQD) is a rough measure of the degree of jointing or fracture in a rock mass, measured as a percentage of the drill core in lengths of 10 cm or more. High-quality rock has an RQD of more than 75%, low quality of less than 50%. Rock quality designation (RQD) has several definitions. The most widely used definition was developed in 1964 by D. U. Deere. It is the borehole core recovery percentage incorporating only pieces of solid core that are longer than 100 mm in length measured along the centerline of the core. In this respect pieces of core that are not hard and sound should not be counted though they are 100 mm in length. RQD was originally introduced for use with core diameters of 54.7 mm (NX-size core). RQD has considerable value in estimating support of rock tunnels. RQD forms a basic element in some of the most used rock mass classification systems: Rock Mass Rating system (RMR) and Q-system.

RQD is defined as the quotient:

$$RQD = \left(\frac{l_{\text{sum of length of core pieces >100mm}}}{l_{\text{tot core run}}} \right) \times 100\%$$

$l_{\text{sum of length of core pieces >10cm}}$ = Sum of length of core pieces that are > 100 mm (4 inches) measured along the centerline.

$l_{\text{tot core run}}$ = Total length of core run.

Classification Table

From the RQD index the rock mass can be classified as follows:

RQD	Rock mass quality
<25%	Very poor
25-50%	Poor
51-75%	Fair
76-90%	Good
91-100%	Excellent

Rock Mass Classification

Many rock mass classification systems use core recovery parameters as input parameter, such as Rock Mass Rating and Q-system.

Crosshole Sonic Logging

Crosshole Sonic Logging (CSL), also known as cross sonic logging or the sonic logging test is the most accurate and reliable technique for assessing the integrity of deep foundation elements such as newly placed concrete drilled shafts and slurry walls, particularly when drilling below groundwater level. Considering the number of issues which can threaten the integrity of deep foundation elements during construction, it is essential to adopt a high level of quality control and assurance. The CSL testing procedure can detect anomalies inside concrete within just minutes.

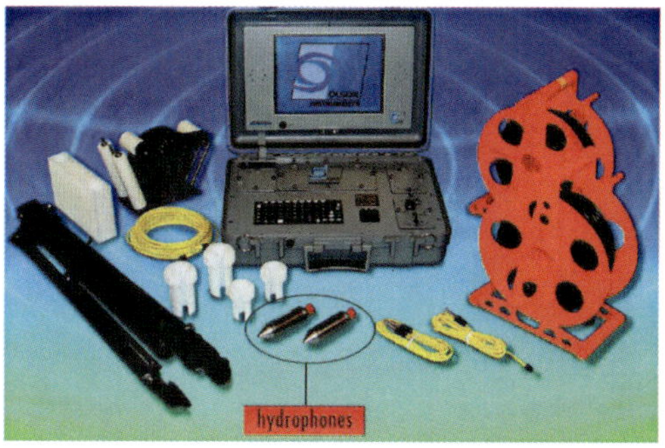

Crosshole Tomography (CT) is a higher resolution analysis which can be used for further investigations after CSL has detected a flaw. Pile tomography delivers unmatched accuracy for defining the size, shape, severity and location of defects. CSL and CT analysis can eliminate the questions and concerns associated with problematic concrete pours.

The Problem

The number of issues which threaten the integrity of drilled shafts and slurry walls are extensive. Necking and caving in of drilled shafts, reinforcing cage buckling, pile damage, tremie 'burping' and excess water intrusion can all have serious effects on structural integrity. Being able to perform QA promptly after the concrete pour can save a substantial amount of time and money compared with late detection of flaws. Having the ability to assess the exact scope of defects will ensure efficient remedial measures.

Applications

CSL and CT can be applied to any structure which supports the transmission of ultrasonic waves. This includes concrete, slurry and grout placed either above or below the waterline. The tests can be applied to concrete drilled shafts, auger cast piles, slurry walls or any mass concrete pour where QA is a priority.

Advantages

- Quick and safe – CSL tests performed in less than 5 minutes.
- Can perform tests for above and below waterline structures.
- Identify and locate the presence of multiple defects.
- CT can deliver high-resolution images with unsurpassed accuracy and interpretation simplicity.
- Perform QA regarding the presence of poor quality concrete, voids, honeycombing and soil intrusion.
- Identify the size, shape, severity, nature and location of anomalies.

How it Works

Installing Access Tubes

Access tubes must be installed before construction and filled with water. These are typically 38 to 50mm diameter PVC or steel tubes tied to the inside of the rebar cage. The number of tubes required depends on the structure's size and the resolution required.

Taking Readings

The source and receiver hydrophones are lowered to the bottom of separate tubes. As the hydrophones are simultaneously lifted, the horizontal ultrasonic pulse velocity is measured. The depth wheel controls the sample reading resolution which is typically taken at 6cm intervals.

Interpreting Results

Uniform, good quality concrete will promote propagation of ultrasonic waves. Longer ultrasonic wave travel times and lower amplitudes correspond to poor quality or voided concrete. A complete loss in signal means a substantial void is hindering the transfer of the signal between the hydrophones.

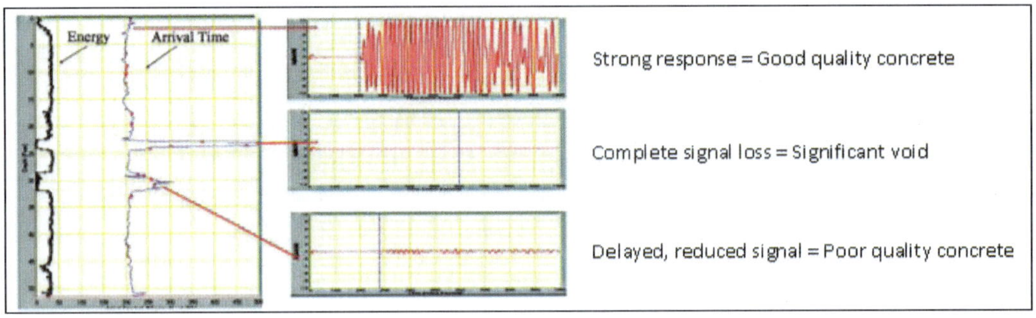

Further Investigation

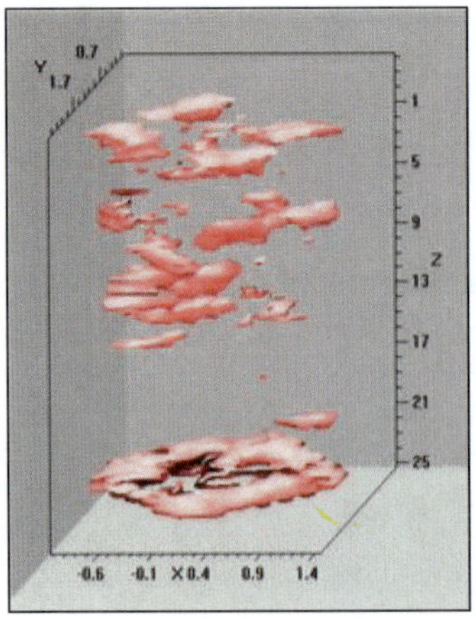

If multiple access tubes are installed, then the perimeter pairs and sub-diagonal tube pairs can be used to further define the 2D image of any anomalies. In addition CT uses differential height measurements to deliver a complete 3D internal image of the structure.

Versatility

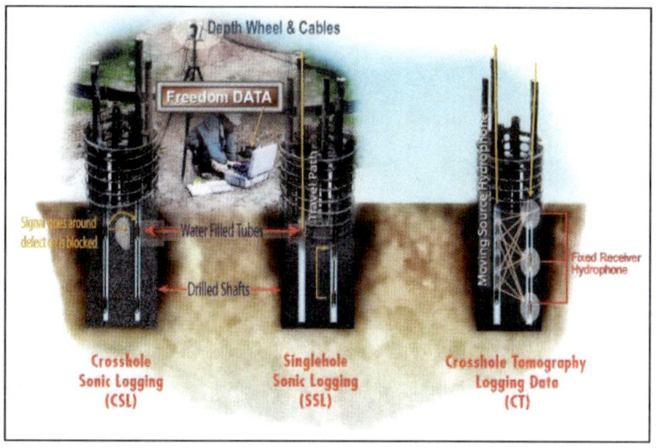

By utilising the same set of water-filled access tubes, both CSL and CT can give invaluable information regarding the quality of concrete between these tubes. Single Hole Sonic Logging (SSL) as presented in the figure to the right, provides additional information regarding concrete quality around the vicinity of an individual tube. This allows QA analysis beyond the perimeter of the rebar cage giving a complete cross-sectional analysis.

Cut - Earthmoving

In civil engineering, a cut or cutting is where soil or rock from a relative rise along a route is removed. The term is also used in river management to speed a waterway's flow by short-cutting a meander.

Cuts are typically used in road, rail, and canal construction to reduce the length and grade of a route. Cut and fill construction uses the spoils from cuts to fill in defiles to cost-effectively create relatively straight routes at steady grades.

Cuts are used as alternatives to indirect routes, embankments, or viaducts. They also have the advantage of comparatively lower noise pollution than elevated or at-grade solutions.

Creation

Open-cut station of the New York City Subway.

Cuts can be created by multiple passes of a shovel, grader, scraper or excavator, or by blasting. One unusual means of creating a cut is to remove the roof of a tunnel through daylighting. Material removed from cuts is ideally balanced by material needed for fills along the same route, but this is not always the case when cut material is unsuitable for use as fill.

The word is also used in the same sense in mining, as in an open-cut mine. The use of cuttings often provides byproducts as a form of mineral extraction, commonly sand, clay or gravel; the cost of building drains, reinforcing banks against landslide and a high water table are factors which commonly limit its use in certain areas.

Types of Cut

There are at least two types of cut, sidehill cut and through cut. The former permits passage of a transportation route alongside of, or around a hill, where the slope is transverse to the roadway or the railway. A sidehill cut can be formed by means of sidecasting, i.e., cutting on the high side balanced by moving the material to build up the low side to achieve a flat surface for the route. In contrast, through cuts, where the adjacent grade is higher on both sides of the route, require removal of material from the area since it cannot be dumped alongside the route. A ledge is a cut in the side of a cliff well above the bottom of a gorge.

Lock Cut

A lock cut on the River, Thames at Bray Lock, Berkshire. The tall wooden poles are designed for boats to tie on to while awaiting entry into the lock.

A lock cut is a section of a river or other inland waterway immediately upstream and downstream of a lock which has been modified to provide locations for boats to moor while waiting for the lock gates to open or to allow people to board or alight vessels.

Cut and Fill

When planning the design and construction of a building, architects and engineers must first consider the existing conditions of the site. Most often, the given site is not level and must be modified before any construction can begin. So, the cut and fill process is typically one of the first construction processes to take place on a site.

Without computer software to determine the amount of land to be added or removed, we can do simple calculations to estimate the cut and fill volume of any site. First, the site is divided into sections. Essentially, the site is "sliced" into segments of land, at a specified fixed distance apart, so we can look at each segment individually. Section cuts are drawn based on the topography of the land in order to accurately represent the land that is being evaluated.

The topography map describes the existing slopes of the land, and allows us to draw site elevations in section cuts.

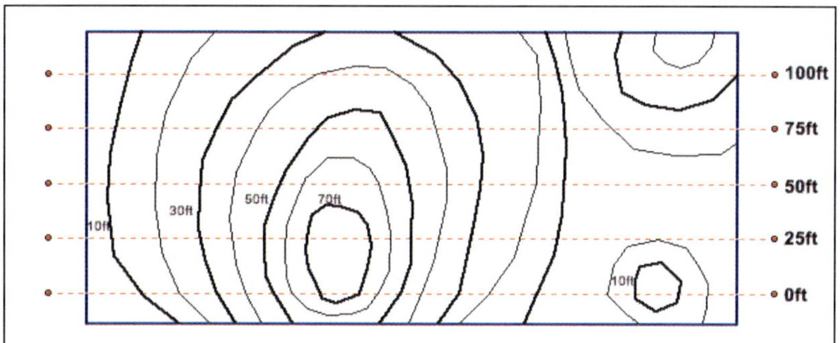

topography map

Each line on the topography map represents a change in elevation of 10 feet. So the section at 0ft (the red dashed line above) looks like this in elevation:

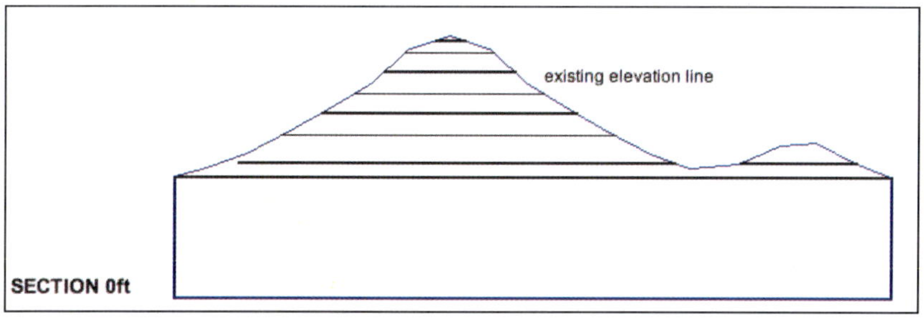

Then, by removing the elevation lines, we get the following section for the existing site elevation:

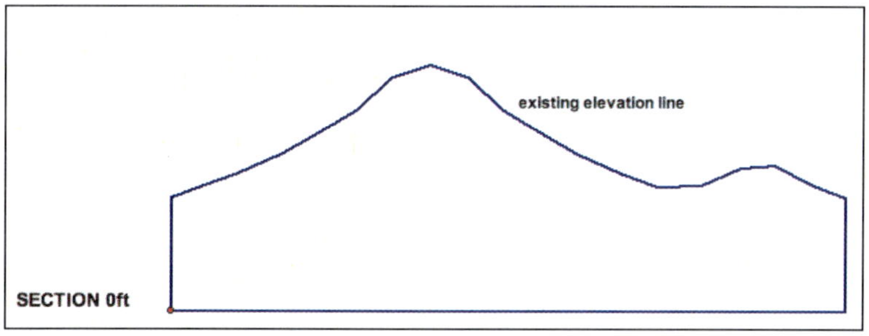

Now that we have two dimensional section cuts, we can estimate the area to be cut and filled in the site. On each section cut the desired land topography is represented as a "cut line" overlaid on the line of the existing land topography. This gives us two lines on every section, and shows us how much of the existing area must be removed or added. Then, the area between these lines is calculated to find the cut and fill areas for each section.

The diagrams below represent the area between the existing elevation line and the cut line. First, we mark the land that must be added to the site to achieve desired land topography. This area is marked "fill."

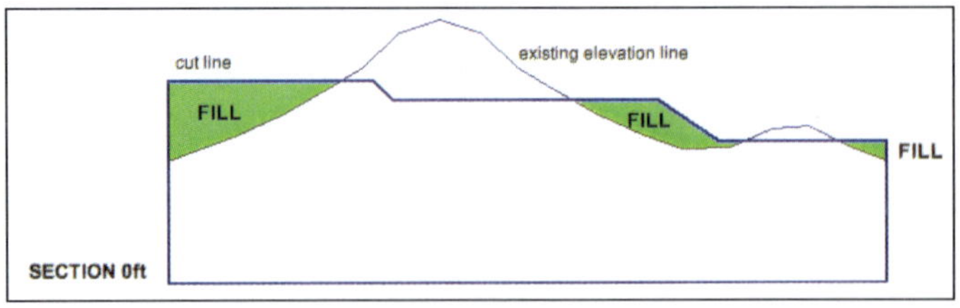

Next, we mark the land that must be removed from the site to achieve desired land topography. This area is marked "cut."

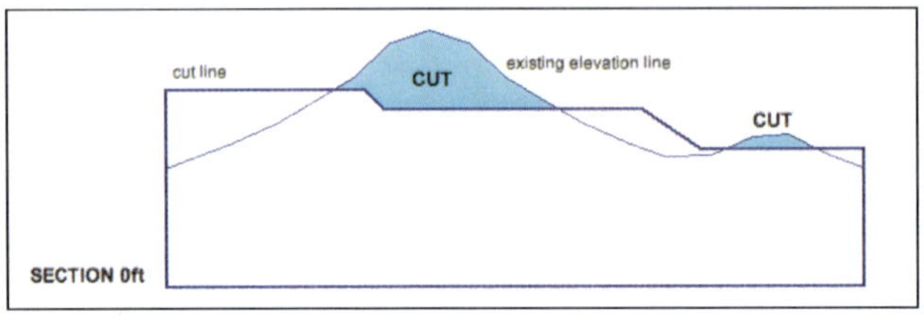

By combining the two diagrams, we can see begin to estimate how much land area must be cut from the site, and how much land area must be filled in the site to level the site for construction.

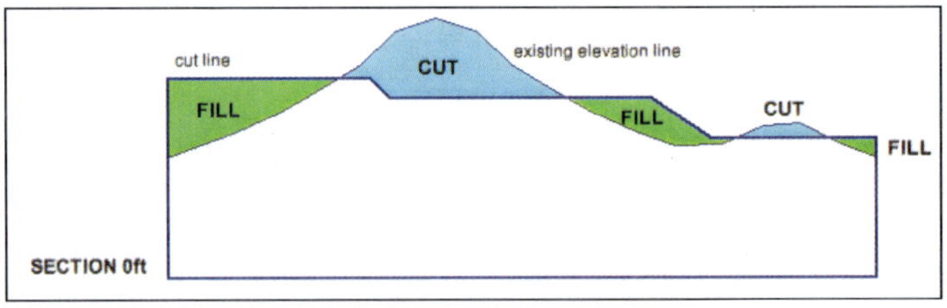

Now, once the section areas have been found, we must account for the additional length of the site (the land between the section cuts) to be able to estimate the volume of the land cut or filled. Finally, after computing the necessary values we are able to determine not only how much land must be removed from or brought to the site, but also where that land must be specifically taken from or brought to. Removing or adding land to any site is a costly process, so it is important that estimations be made as close to the existing conditions as possible. Once we know how much land to cut or fill, and where to cut and fill it, work can begin to properly level the site and construct building.

Site Area

The following activity uses the mathematical concepts of estimating area and volume to solve a realistic cut and fill problem. Each two-dimensional section cut is overlaid with a grid to allow us to estimate the area between the existing elevation line and the cut line.

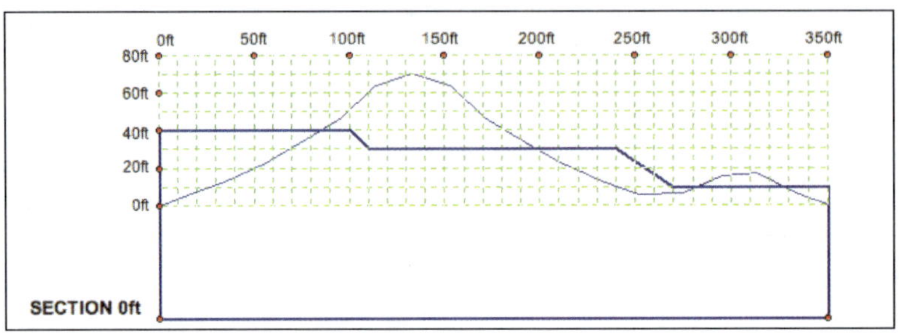

To estimate area, we can divide the area under the curve into rectangles and then find the area of each rectangle. Refer to the diagram below for the process of drawing rectangles. Consistency in drawing the rectangles is very important to ensure proper balance between over and under estimation of area. Since all of the intervals are the same width of 10ft, the only estimating that you will need to do is with the height of each rectangle. For this exercise estimate the height to the nearest 2 1/2 feet (or 1/4 of the 10' X 10' square).

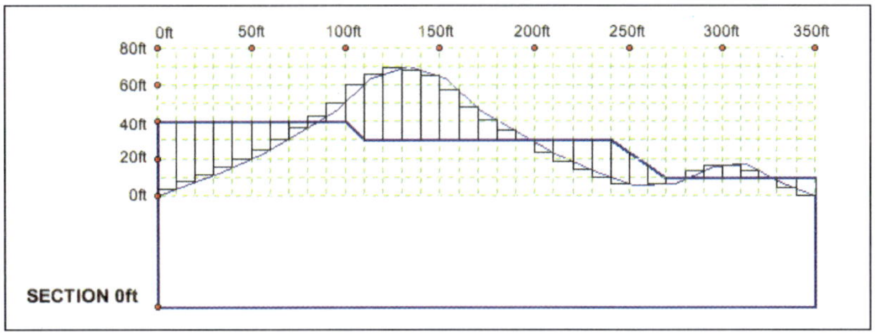

By removing the grid lines, it is easy to see the areas of the rectangles that we will be estimating. Since these rectangles will represent and overestimation in some parts, and an underestimation in others, we will consistently draw rectangles and not worry about the area between the lines that are missed.

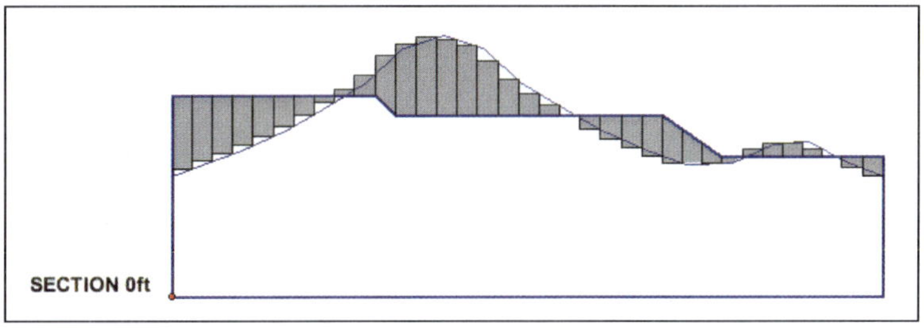

The diagrams below represent two different ways rectangles can be drawn between the existing elevation line and the cut line. The process of left and right estimation indicate the direction in which the rectangles that meet the curve and are drawn.

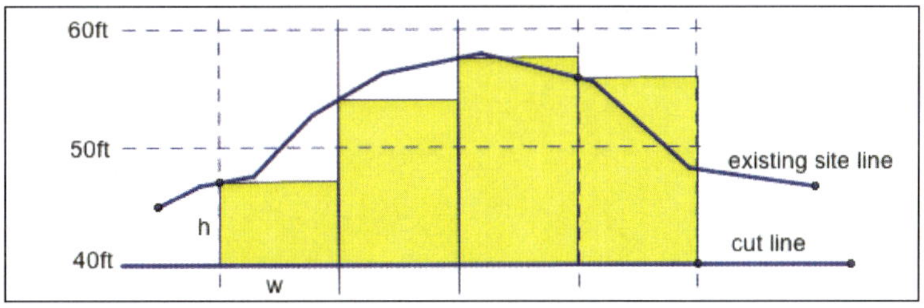

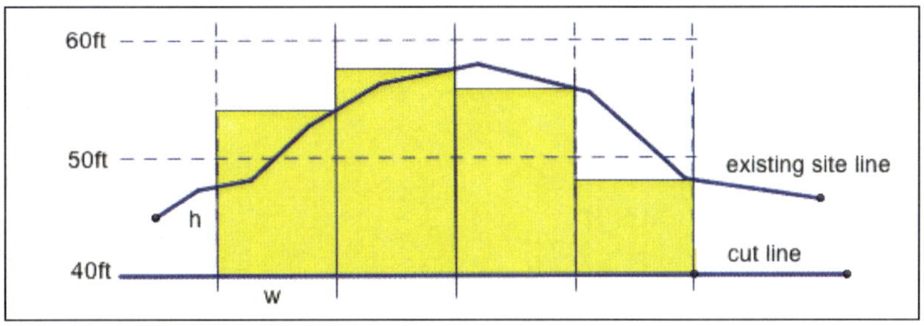

Site Volume

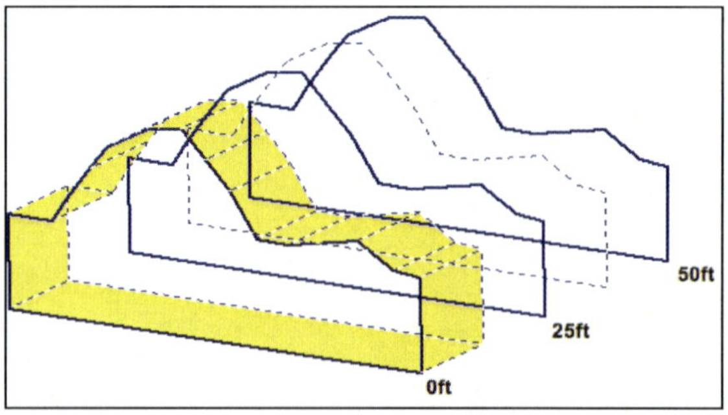

To estimate the volume, we use the area that has been determined (as width and height) and then multiply by the distance between each section (depth). Note that the first and last section is on the site boundary. Notice also that these boundary sections only have a volume on one of their sides. The remaining sections between the boundaries have land volume on each side. See the diagram below. The solid lines refer to section cuts, and the dashed lines refer to the midpoint between the section cuts. The volume for section cut 0ft is shaded below.

References

- The CNN Wire Staff (2010-06-03). "'Top Kill' fails, BP moves on 'to next option.'". Cnn.com. Retrieved 2010-06-23
- 42482-fundamentals-of-geotechnical-engineering, geotechnical-engineering: brighthubengineering.com, Retrieved 16 January, 2019
- Mitchell, J.K., and Soga, K. (2005) Fundamentals of soil behavior, Third edition, John Wiley and Sons, Inc., ISBN 978-0-471-46302-3
- Soil-gradation: wikimili.com, Retrieved 30 August, 2019
- Santamarina, J.C., Klein, K.A., & Fam, M.A. (2001). Soils and Waves: Particulate Materials Behavior, Characterization and Process Monitoring. Wiley. ISBN 978-0-471-49058-6
- "Mine Safety and Health Administration (MSHA) - Interactive Training - Surface Powered Haulage Safety". Msha.gov. Retrieved 2010-06-23
- Crosshole-sonic-logging-and-crosshole-tomography: pcte.com.au, Retrieved 29 March, 2019
- Mccarthy, David F. (2007). Essentials of Soil Mechanics and Foundations. Upper Saddle River, NJ: Pearson Prentice Hall. P. 595. ISBN 0-13-114560-6

5

Earthquake Engineering

The interdisciplinary branch of engineering that analyzes and designs structures like buildings and bridges with regard to earthquakes is known as earthquake engineering. It's used to design earthquake-resistant structures and implement active vibration control. This chapter has been carefully written to provide an easy understanding of these applications of earth-quake engineering.

Earthquake engineering is an interdisciplinary branch of engineering that designs and analyzes structures, such as buildings and bridges, with earthquakes in mind. Its overall goal is to make such structures more resistant to earthquakes. An earthquake (or seismic) engineer aims to construct structures that will not be damaged in minor shaking and will avoid serious damage or collapse in a major earthquake. Earthquake engineering is the scientific field concerned with protecting society, the natural environment, and the man-made environment from earthquakes by limiting the seismic risk to socio-economically acceptable levels. Traditionally, it has been narrowly defined as the study of the behavior of structures and geo-structures subject to seismic loading; it is considered as a subset of structural engineering, geotechnical engineering, mechanical engineering, chemical engineering, applied physics, etc. However, the tremendous costs experienced in recent earthquakes have led to an expansion of its scope to encompass disciplines from the wider field of civil engineering, mechanical engineering and from the social sciences, especially sociology, political science, economics and finance.

An earthquake is a sudden slipping or movement of a portion of the Earth's crust or plates, caused by a sudden release of stresses. Earthquake epicenters are usually less than 25 miles below the Earth's surface and are accompanied and followed by a series of vibrations.

What causes Earthquakes and where do Earthquakes Happen?

The earth has four major layers: The inner core, outer core, mantle and crust. The crust and the top of the mantle make up a thin layer on the surface of earth. But this layer is not a single cover, it is made up of many pieces like jigsaw covering the surface of the earth. These keep slowly moving around each other, slidepast one another and bump into each other. These puzzle pieces are called tectonic plates, and the edges of the plates are called the plate boundaries. The plate boundaries are made up of many faults, and most of the earthquakes around the world occur on these faults. Since the edges of the plates are rough, they get stuck while the rest of the plate keeps moving. Finally, when the plate has moved far enough, the edges unstick on one of the faults and there is an earthquake.

Types of Earthquakes

Most earthquakes in the world occur along the boundaries of the tectonic plates and are called Inter-plate Earthquakes. A number of earthquakes also occur within the plate itself away from the plate boundaries, called Intra-plate Earthquakes.

How are Earthquakes Recorded?

Earthquakes are recorded by instrument called seismographs. The recording they made, is called a seismogram. The seismo gram consists of two parts, a base and a weight, to held it firmly in the ground. When an earthquake causes the ground to shake, the base of the seismograph shakes too, but the hanging weight does not. Instead the spring or string that it is hanging from absorbs all the movement. Thus the difference between the moving and immovable part is recorded.

How are Earthquakes Measured?

The size of an earthquake depends on the size of the fault and the amount of slip on the fault, but this cannot be measured directly as faults are deep in the earth. The seismogram recordings made on the seismographs at the surface of the earth are used to determine the intensity of earthquake. A short line with less zigzag portions represents a small earthquake and a lengthy line with a lot of zigzag sections shows a large earthquake. The length of line on the seismograph depends on the size of the fault and the wigginess of the line depends upon the amount of slip of the fault. The size or intensity of earthquake is called Magnitude of earthquake.

Aims and Objectives of Earthquake Engineering Study

The main objective of this volume is to illustrate to students of structural and architectural engineering the problems and solutions in attaining efficient earthquake-resistant structures and facilities. To achieve this objective, after a brief discussion of the general goals in seismic-resistant design and construction of structures and facilities, the different sources of damage that can be triggered by an earthquake are discussed and illustrated.

Emphasis is placed on the discussion and illustration of damage induced by vibration on timber, masonry, concrete and steel structures. The importance of a comprehensive approach to the problem of earthquake resistant construction is emphasized next and the need for placing more emphasis on conceptual design is discussed by offering guidelines for and illustrations of efficient seismic-resistant design. The need for research in earthquake-resistant design and construction is briefly discussed and examples of integrated experimental and analytical investigations in the development of modern seismic-resistant design are also shown.

Earthquake-resistant Structures

Earthquake-resistant structures are structures designed to protect buildings from earthquakes. While no structure can be entirely immune to damage from earthquakes, the goal of earthquake-resistant construction is to erect structures that fare better during seismic activity than their conventional counterparts. According to building codes, earthquake-resistant structures are intended to withstand the largest earthquake of a certain probability that is likely to occur at their location.This means the loss of life should be minimized by preventing collapse of the buildings for rare earthquakes while the loss of the functionality should be limited for more frequent ones.

To combat earthquake destruction, the only method available to ancient architects was to build their landmark structures to last, often by making them excessively stiff and strong.

Currently, there are several design philosophies in earthquake engineering, making use of experimental results, computer simulations and observations from past earthquakes to offer the required performance for the seismic threat at the site of interest. These range from appropriately sizing the structure to be strong and ductile enough to survive the shaking with an acceptable damage, to equipping it with base isolation or using structural vibration control technologies to minimize any forces and deformations. While the former is the method typically applied in most earthquake-resistant structures, important facilities, landmarks and cultural heritage buildings use the more advanced (and expensive) techniques of isolation or control to survive strong shaking with minimal damage. Examples of such applications are the Cathedral of Our Lady of the Angels and the Acropolis Museum.

Trends and Projects

Some of the new trends and projects in the field of earthquake engineering structures are presented.

Building Materials

Based on studies in New Zealand, relating to Christchurch earthquakes, precast concrete designed and installed in accordance with modern codes performed well. According to the Earthquake Engineering Research Institute, precast panel buildings had good durability during the earthquake in Armenia, compared to precast frame-panels.

Earthquake Shelter

One Japanese construction company has developed a six-foot cubical shelter, presented as an alternative to earthquake-proofing an entire building.

Concurrent Shake-table Testing

Concurrent shake-table testing of two or more building models is a vivid, persuasive and effective way to validate earthquake engineering solutions experimentally.

Thus, two wooden houses built before adoption of the 1981 Japanese Building Code were moved to E-Defense for testing. The left house was reinforced to enhance its seismic resistance, while the other one was not. These two models were set on E-Defense platform and tested simultaneously.

Combined Vibration Control Solution

Designed by architect Merrill W. Baird of Glendale, working in collaboration with A. C. Martin

Architects of Los Angeles, the Municipal Services Building at 633 East Broadway, Glendale was completed in 1966. Prominently sited at the corner of East Broadway and Glendale Avenue, this civic building serves as a heraldic element of Glendale's civic center.

Close-up of abutment of seismically retrofitted Municipal Services Building in Glendale, California.

Seismically retrofitted Municipal Services Building in Glendale.

In October 2004 Architectural Resources Group (ARG) was contracted by Nabih Youssef & Associates, Structural Engineers, to provide services regarding a historic resource assessment of the building due to a proposed seismic retrofit.

In 2008, the Municipal Services Building of the City of Glendale, California was seismically retrofitted using an innovative combined vibration control solution: the existing elevated building foundation of the building was put on high damping rubber bearings.

Steel Plate Walls System

Coupled steel plate shear walls, Seattle.

A steel plate shear wall (SPSW) consists of steel infill plates bounded by a column-beam system. When such infill plates occupy each level within a framed bay of a structure, they constitute a SPSW system. Whereas most earthquake resistant construction methods are adapted from older systems, SPSW was invented entirely to withstand seismic activity.

The Ritz-Carlton/JW Marriott hotel building engaging the advanced steel plate shear walls system, Los Angeles.

SPSW behavior is analogous to a vertical plate girder cantilevered from its base. Similar to plate girders, the SPSW system optimizes component performance by taking advantage of the post-buckling behavior of the steel infill panels.

The Ritz-Carlton/JW Marriott hotel building, a part of the LA Live development in Los Angeles, California, is the first building in Los Angeles that uses an advanced steel plate shear wall system to resist the lateral loads of strong earthquakes and winds.

Kashiwazaki-Kariwa Nuclear Power Plant is Partially Upgraded

The Kashiwazaki-Kariwa Nuclear Power Plant, the largest nuclear generating station in the world by net electrical power rating, happened to be near the epicenter of the strongest M_w 6.6 July 2007 Chūetsu offshore earthquake. This initiated an extended shutdown for structural inspection which indicated that a greater earthquake-proofing was needed before operation could be resumed.

On May 9, 2009, one unit (Unit 7) was restarted, after the seismic upgrades. The test run had to continue for 50 days. The plant had been completely shut down for almost 22 months following the earthquake.

Seismic Test of Seven-story Building

A destructive earthquake struck a lone, wooden condominium in Japan. The experiment was webcast live on July 14, 2009 to yield insight on how to make wooden structures stronger and better able to withstand major earthquakes.

The Miki shake at the Hyogo Earthquake Engineering Research Center is the capstone experiment of the four-year NEESWood project, which receives its primary support from the U.S. National Science Foundation Network for Earthquake Engineering Simulation (NEES) Program.

"NEESWood aims to develop a new seismic design philosophy that will provide the necessary mechanisms to safely increase the height of wood-frame structures in active seismic zones of the United States, as well as mitigate earthquake damage to low-rise wood-frame structures," said Rosowsky, Department of Civil Engineering at Texas A&M University. This philosophy is based on the application of seismic damping systems for wooden buildings. The systems, which can be installed inside the walls of most wooden buildings, include strong metal frame, bracing and dampers filled with viscous fluid.

Superframe Earthquake Proof Structure

The proposed system is composed of core walls, hat beams incorporated into the top level, outer columns and viscous dampers vertically installed between the tips of the hat beams and the outer columns. During an earthquake, the hat beams and outer columns act as outriggers and reduce the overturning moment in the core, and the installed dampers also reduce the moment and the lateral deflection of the structure. This innovative system can eliminate inner beams and inner columns on each floor, and thereby provide buildings with column-free floor space even in highly seismic regions.

Active Vibration Control

Active vibration control is the active application of force in an equal and opposite fashion to the forces imposed by external vibration. With this application, a precision industrial process can be maintained on a platform essentially vibration-free.

Many precision industrial processes cannot take place if the machinery is being affected by vibration. For example, the production of semiconductor wafers requires that the machines used for the photolithography steps be used in an essentially vibration-free environment or the sub-micrometre features will be blurred. Active vibration control is now also commercially available for reducing vibration in helicopters, offering better comfort with less weight than traditional passive technologies.

In the past, passive techniques were used. These include traditional vibration dampers, shock absorbers, and base isolation.

The typical active vibration control system uses several components:

- A massive platform suspended by several active drivers (that may use voice coils, hydraulics, pneumatics, piezo-electric or other techniques).
- Three accelerometers that measure acceleration in the three degrees of freedom.
- An electronic amplifier system that amplifies and inverts the signals from the accelerometers. A PID controller can be used to get better performance than a simple inverting amplifier.
- For very large systems, pneumatic or hydraulic components that provide the high drive power required.

If the vibration is periodic, then the control system may adapt to the ongoing vibration, thereby providing better cancellation than would have been provided simply by reacting to each new acceleration without referring to past accelerations.

Active vibration control strategy has been successfully implemented for vibration attenuation of beam, plate and shell structures by numerous researchers. For active vibration control strategy, the structure should be smart enough to sense external disturbances and react accordingly. In order to develop an active structure (also known as smart structure), smart materials must be integrated or embedded with the structure. The smart structure involves sensors (strain, acceleration, velocity, force etc.), actuators (force, inertial, strain etc.) and a control algorithm (feedback or feed forward). The number of smart materials have been investigated and fabricated over the years; some of them are shape memory alloys, piezoelectric materials, optical fibers, electro-rheological fluids, magneto-strictive materials.

Endurance Time Method

Endurance time (ET) method is a dynamic structural analysis procedure for seismic assessment of structures. In this procedure, an intensifying dynamic excitation is used as the loading function. Endurance time method is a time-history based dynamic analysis procedure. An estimate of the structural response at different equivalent seismic intensity levels is obtained in a single response history analysis. This method has applications in seismic assessment of various structural types and in different areas of earthquake engineering.

The Concept of Endurance Time Method

Endurance time (ET) method is a dynamic structural analysis procedure in which intensifying dynamic excitation is used as the loading function. An estimate of structural response and/or performance at the entire seismic intensity range of interest is obtained in each response history analysis. The concept of endurance time analysis is similar to the exercise test applied in medicine. Similar concept has also been extended to applications in the analysis of offshore platforms under water waves.

ET Excitation Functions

ET excitation functions are generated by using numerical optimization methods. ET excitation functions are publicly available through internet websites. ET excitation functions are categorized into four generations as follows:

- First generation of ET excitation functions (ETEFs) are essentially a filtered and profiled white noise. These were used for demonstrating the concept of ET and have limited practical significance.
- Second-generation ETEFs incorporate response spectrum matching. These ETEFs produce numerically significant analysis results.
- Third-generation ETEFs are optimized in nonlinear range. These ETEFs deliver improved analysis performance.
- Fourth-generation ETEFs are optimized to include duration consistency.

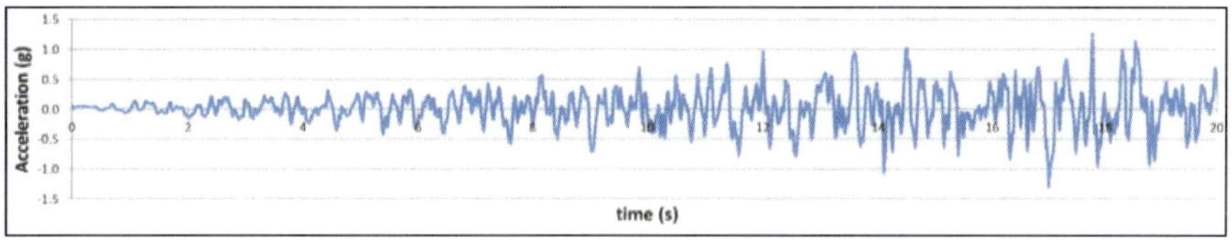

A typical endurance time excitation function.

Application Areas in Earthquake Engineering

Endurance time method has been applied in the following areas of earthquake engineering:

- Nonlinear dynamic analysis of structures.
- Seismic evaluation of jacket-type offshore platforms.
- Optimal damper placement in framed buildings.
- Seismic assessment of structures.
- Performance-based design.
- Structural optimization.
- Multi-component seismic analysis.
- Soil–structure interaction.
- soil-pile-superstructure interaction.
- Liquid–structure interaction.
- Dam engineering.
- Bridge engineering.
- Seismic rehabilitation.
- Collapse analysis.

Structural Type Applications

ET method has been applied in seismic assessment of the following structural types:

- Single degree of freedom systems.
- Moment and braced steel frames.
- Concrete frames.
- Bridges.
- Gravity dams.
- Arch dams.
- Shell structures.
- Steel tanks.
- Offshore structures.

Advantages of ET Method

Major advantages of the endurance time method are as follows:

- ET significantly reduces the computational demand required for performing a standard response history analysis of structures for seismic assessment, especially when response at multiple levels of intensity is to be considered.

- ET is applicable in a wide range of seismic assessment problems and provides a generic approach for the seismic analysis of a wide range of structural types.
- ET method is reasonably simple and sensible when a realistic dynamic analysis of a complex structure is required.

Limitations of ET Method

Major limitations of the endurance time method are as follows:

- ET is an approximate method for predicting the structural response.
- The production of usable ETEFs that are applicable in a particular situation can be complicated.
- The procedure is still under development and sufficient background information may not be available for specific applications.

Sarma Method

The Sarma method is a method used primarily to assess the stability of soil slopes under seismic conditions. Using appropriate assumptions the method can also be employed for static slope stability analysis. It was proposed by Sarada K. Sarma in the early 1970s as an improvement over the other conventional methods of analysis which had adopted numerous simplifying assumptions.

Method

Assumptions

The method satisfies all conditions of equilibrium, (i.e. horizontal and vertical force equilibrium and moment equilibrium for each slice). It may be applied to any shape of slip surface as the slip surfaces are not assumed to be vertical, but they may be inclined. It is assumed that magnitudes of vertical side forces follow prescribed patterns. For n slices (or wedges), there are 3n equations and 3n unknowns, and therefore it statically determinate without the need of any further additional assumptions.

Advantages

The Sarma method is called an advanced and rigorous method of static and seismic slope stability analysis. It is called advanced because it can take account of non-circular failure surfaces. Also, the multi-wedge approach allows for non-vertical slices and irregular slope geometry. It is called a rigorous method because it can satisfy all the three conditions of equilibrium, horizontal and vertical forces and moments. The Sarma method is nowadays used as a verification to finite element programs (also FE limit analysis) and it is the standard method used for seismic analysis.

Use

The method is used mainly for two purposes, to analyse earth slopes and earth dams. When used to analyse seismic slope stability it can provide the factor of safety against failure for a given earthquake load, i.e. horizontal seismic force ir acceleration (critical acceleration). Besides, it can provide the required earthquake load (force or acceleration) for which a given slope will fail, i.e. the factor of safety will be equal to 1.

When the method is used in the analysis of earth dams (i.e. the slopes of the dam faces), the results of the analysis, i.e. the critical acceleration is used in the Newmark's sliding block analysis in order to calculate the induced permanent displacements. This follows the assumption that displacements will result if the earthquake induced accelerations exceed the value of the critical acceleration for stability.

Accuracy

General Acceptance

The Sarma method has been extensively used in seismic analysis software for many years and has been the standard practice until recently for seismic slope stability for many years (similar to the Mononobe–Okabe method for retaining walls). Its accuracy has been verified by various researchers and it has been proved to yield results quite similar to the modern safe Lower Bound numerical stability Limit Analysis methods.

Modern Alternatives

However, nowadays modern numerical analysis software employing usually the finite element, finite difference and boundary element methods are more widely used for special case studies. Particular attention has been recently given to the finite element method which can provide very accurate results through the release of several assumptions usually adopted by the conventional methods of analysis. Special boundary conditions and constitutive laws can model the case in a more realistic fashion.

Seismic Loading

The Seismic Loading option allows you to include a pseudo-static seismic (earthquake) loading in the finite element analysis. To define seismic loading:

Select Seismic Loading from the Loading menu. Note: you can also right-click on the Stress Block/ Seismic Load icon and select Seismic Loading from the popup menu.

In the Seismic Load dialog, enter seismic coefficients for the Horizontal and Vertical directions. You may also select the stage at which the seismic force is applied. Select OK.

Seismic Coefficients

The Seismic Coefficients are dimensionless coefficients which represent the (maximum) earthquake acceleration as a fraction of the acceleration due to gravity. Typical values are in the range of 0.1 to 0.3.

When a seismic coefficient is defined, an additional Body Force will be applied to each finite element in the mesh, as follows:

Seismic Force = Seismic Coefficient * Body Force (due to gravity)

= Seismic Coefficient * (area of element * Unit Weight of element material)

- Body Force (due to gravity) is simply the self-weight of a finite element.
- The Seismic body force is vectorially added to the (downward) Body Force which exists due to gravity, to obtain the total body force acting on the element.

Seismic Force and Initial Element Loading

It is very important to note that Seismic Force only applies to materials which have an Initial Element Loading which includes Body Force. Initial Element Loading is specified in the Define Material Properties dialog.

If Body Force is not included in the Initial Element Loading, then the applied Seismic Force will be zero, even if seismic coefficients are defined in the Seismic Load dialog. You must remember this if you want to use Seismic Load in combination with Constant Field Stress, since by default, the Initial Element Loading for Constant Field Stress is Field Stress Only.

Direction Convention

The following direction convention is used for the seismic coefficients in RS2:

- A POSITIVE Horizontal coefficient represents a force applied to the RIGHT.
- A NEGATIVE Horizontal coefficient represents a force applied to the LEFT.
- A POSITIVE Vertical coefficient represents a force applied UPWARDS.
- A NEGATIVE Vertical coefficient represents a force applied DOWNWARD.

When carrying out a finite element analysis with Seismic Loading, you should run multiple analyses using different combinations and directions of Horizontal and Vertical seismic coefficients, in order to evaluate the effect of seismic force direction on the analysis results (e.g. stresses, displacements, strength factor etc).

Application of Seismic Coefficient

The Application of Seismic Coefficient option allows you to specify the stage at which the seismic load is applied. This is only applicable for multi-stage models (i.e. Number of Stages > 1).

Applied in Stage

The Applied in Stage option allows you to select the stage at which the seismic load will be applied. Strictly speaking, the seismic load should be applied at the final stage of excavation. If seismic load is applied at an intermediate stage of excavation, the effect of the load will propagate to all subsequent stages (i.e. the seismic load is not removed at subsequent stages, after the initial stage of application).

Applied when Body Force is Applied

The Applied when body force is applied option will automatically apply the seismic load at the stage when the body force of an element (material) is first applied. Generally this will be the first stage. Furthermore, the seismic load will be applied according to the load split, if the Load Split option is in use. This means that the seismic load will be applied in successive stages, according to the staging of the load split factors which are applied to the body force.

This option only exists to maintain backward compatibility with RS2 version 4.0. The use of this option is not recommended.

Seismic Load Icon

When (non-zero) seismic coefficients are entered in the Seismic Loading dialog, a Seismic Load icon will be displayed in the upper right corner of the view. The icon indicates that seismic load has been applied to the model, and the magnitude and direction of the seismic coefficient(s).

Seismic load icon (horizontal coefficient = 0.2, vertical coefficient = -0.1)

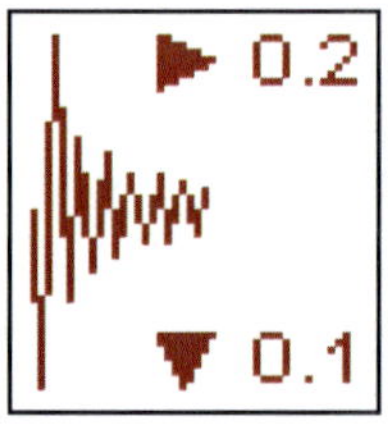

Seismic Analysis

Seismic analysis is a subset of structural analysis and is the calculation of the response of a building (or nonbuilding) structure to earthquakes. It is part of the process of structural design, earthquake engineering or structural assessment and retrofit in regions where earthquakes are prevalent.

As seen in the figure, a building has the potential to 'wave' back and forth during an earthquake (or even a severe wind storm). This is called the 'fundamental mode', and is the lowest frequency of building response. Most buildings, however, have higher modes of response, which are uniquely activated during earthquakes. The figure just shows the second mode, but there are higher 'shimmy' (abnormal vibration) modes. Nevertheless, the first and second modes tend to cause the most damage in most cases.

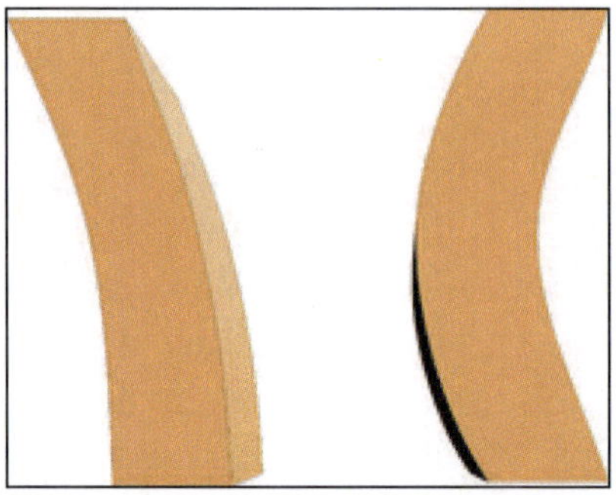

First and second modes of building seismic response.

The earliest provisions for seismic resistance were the requirement to design for a lateral force equal to a proportion of the building weight (applied at each floor level). This approach was adopted in the appendix of the 1927 Uniform Building Code (UBC), which was used on the west coast of the United States. It later became clear that the dynamic properties of the structure affected the loads generated during an earthquake. In the Los Angeles County Building Code of 1943 a provision to vary the load based on the number of floor levels was adopted. The concept of "response spectra" was developed in the 1930s, but it wasn't until 1952 that a joint committee of the San Francisco Section of the ASCE and the Structural Engineers Association of Northern California (SEAONC) proposed using the building period (the inverse of the frequency) to determine lateral forces.

The University of California, Berkeley was an early base for computer-based seismic analysis of structures, led by Professor Ray Clough (who coined the term finite element). Students included Ed Wilson, who went on to write the program SAP in 1970, an early "finite element analysis" program.

Earthquake engineering has developed a lot since the early days, and some of the more complex designs now use special earthquake protective elements either just in the foundation (base isolation) or distributed throughout the structure. Analyzing these types of structures requires specialized explicit finite element computer code, which divides time into very small slices and models the actual physics, much like common video games often have "physics engines". Very large and complex buildings can be modeled in this way (such as the Osaka International Convention Center).

Structural analysis methods can be divided into the following five categories:

Equivalent Static Analysis: This approach defines a series of forces acting on a building to represent the effect of earthquake ground motion, typically defined by a seismic design response spectrum. It assumes that the building responds in its fundamental mode. For this to be true, the building must be low-rise and must not twist significantly when the ground moves. The response is read from a design response spectrum, given the natural frequency of the building (either calculated or defined by the building code). The applicability of this method is extended in many building codes by applying factors to account for higher buildings with some higher modes, and for low levels of twisting. To account for effects due to "yielding" of the structure, many codes apply modification factors that reduce the design forces (e.g. force reduction factors).

Response Spectrum Analysis: This approach permits the multiple modes of response of a building to be taken into account (in the frequency domain). This is required in many building codes for all except very simple or very complex structures. The response of a structure can be defined as a combination of many special shapes (modes) that in a vibrating string correspond to the "harmonics". Computer analysis can be used to determine these modes for a structure. For each mode, a response is read from the design spectrum, based on the modal frequency and the modal mass, and they are then combined to provide an estimate of the total response of the structure. In this we have to calculate the magnitude of forces in all directions i.e. X, Y & Z and then see the effects on the building. Combination methods include the following:

- Absolute – peak values are added together.
- Square root of the sum of the squares (SRSS).

- Complete quadratic combination (CQC) – a method that is an improvement on SRSS for closely spaced modes.

The result of a response spectrum analysis using the response spectrum from a ground motion is typically different from that which would be calculated directly from a linear dynamic analysis using that ground motion directly, since phase information is lost in the process of generating the response spectrum.

In cases where structures are either too irregular, too tall or of significance to a community in disaster response, the response spectrum approach is no longer appropriate, and more complex analysis is often required, such as non-linear static analysis or dynamic analysis.

Linear Dynamic Analysis: Static procedures are appropriate when higher mode effects are not significant. This is generally true for short, regular buildings. Therefore, for tall buildings, buildings with torsional irregularities, or non-orthogonal systems, a dynamic procedure is required. In the linear dynamic procedure, the building is modelled as a multi-degree-of-freedom (MDOF) system with a linear elastic stiffness matrix and an equivalent viscous damping matrix.

The seismic input is modelled using either modal spectral analysis or time history analysis but in both cases, the corresponding internal forces and displacements are determined using linear elastic analysis. The advantage of these linear dynamic procedures with respect to linear static procedures is that higher modes can be considered. However, they are based on linear elastic response and hence the applicability decreases with increasing nonlinear behaviour, which is approximated by global force reduction factors.

In linear dynamic analysis, the response of the structure to ground motion is calculated in the time domain, and all phase information is therefore maintained. Only linear properties are assumed. The analytical method can use modal decomposition as a means of reducing the degrees of freedom in the analysis.

Nonlinear Static Analysis: In general, linear procedures are applicable when the structure is expected to remain nearly elastic for the level of ground motion or when the design results in nearly uniform distribution of nonlinear response throughout the structure. As the performance objective of the structure implies greater inelastic demands, the uncertainty with linear procedures increases to a point that requires a high level of conservatism in demand assumptions and acceptability criteria to avoid unintended performance. Therefore, procedures incorporating inelastic analysis can reduce the uncertainty and conservatism.

This approach is also known as "pushover" analysis. A pattern of forces is applied to a structural model that includes non-linear properties (such as steel yield), and the total force is plotted against a reference displacement to define a capacity curve. This can then be combined with a demand curve (typically in the form of an acceleration-displacement response spectrum (ADRS)). This essentially reduces the problem to a single degree of freedom (SDOF) system.

Nonlinear static procedures use equivalent SDOF structural models and represent seismic ground motion with response spectra. Story drifts and component actions are related subsequently to the global demand parameter by the pushover or capacity curves that are the basis of the non-linear static procedures.

Nonlinear Dynamic Analysis: Nonlinear dynamic analysis utilizes the combination of ground motion records with a detailed structural model, therefore is capable of producing results with relatively low uncertainty. In nonlinear dynamic analyses, the detailed structural model subjected to a ground-motion record produces estimates of component deformations for each degree of freedom in the model and the modal responses are combined using schemes such as the square-root-sum-of-squares.

In non-linear dynamic analysis, the non-linear properties of the structure are considered as part of a time domain analysis. This approach is the most rigorous, and is required by some building codes for buildings of unusual configuration or of special importance. However, the calculated response can be very sensitive to the characteristics of the individual ground motion used as seismic input; therefore, several analyses are required using different ground motion records to achieve a reliable estimation of the probabilistic distribution of structural response. Since the properties of the seismic response depend on the intensity, or severity, of the seismic shaking, a comprehensive assessment calls for numerous nonlinear dynamic analyses at various levels of intensity to represent different possible earthquake scenarios. This has led to the emergence of methods like the incremental dynamic analysis.

Seismic Code

Seismic codes are a sub-division of the national building codes, the implementation of which can be divided into immediate-, medium- and long-term activities. Since the implementation of seismic codes will affect all new and existing buildings, the modification of existing codes and standards requires local legislation and extensive capacity building of the controlling and implementing agencies (private, institutional and government). To upgrade the existing building stock to the level of the new (better) codes and standards is a massive undertaking in labour and finances, requiring a thorough analysis of the existing building stock and the development of plans for retrofitting of millions of buildings.

Effectiveness of Seismic Codes

"Earthquakes Don't Kill People; Buildings Do" is a very well known quote, used in many reports about earthquake damage and preparedness materials. Building codes in general are designed to provide quality and safety for the building occupants and safeguard their assets and economic investments. Seismic codes are designed to prevent the total collapse of the building during a maximum estimated earthquake event. The general seismic codes need to be complemented with national data about the strength and characteristics of that maximum earthquake event per zone, since the strength of the earthquake will decrease with the distance from the epicentre. The PGAg, soil conditions, resistance of the buildings and the number of buildings per region will determine the total economic loss after each event.

Several points about the importance of building codes have been made by these and other seismic experts:

- With the implementation of seismic codes the level of damage would be far less.

- Buildings that predate the adoption of the codes did not comply with the code.
- In rural areas buildings are not designed by engineers or architects, but rather constructed with local experience.
- Un-reinforced masonry buildings are very vulnerable to earthquakes.
- New buildings do not necessarily comply with all seismic code regulations.

Upgrading or Retrofitting Building Stock that Predates Seismic Codes

The main financial problem lies in upgrading old buildings (ie. those that predate seismic codes) to meet seismic standards, which often requires a far greater percentage of the building value as opposed to adding features to a new construction. In the case of low-rise rural houses, it is usually more cost-efficient to pull down the old building and construct an entirely new building. This, however, requires a major investment for low-income families. External or community-based control on extensions and new buildings would ensure that seismic codes are correctly implemented. In LICs when subsidies are unavailable, income generating projects, in combination with education and community-based construction control, are potential ways to realise new buildings that adhere to the codes.

Thus for LICs it is very important to comply with seismic codes, as not doing so negatively affects the resilience of the entire population.

References

- Valamanesh, V and Estekanchi, H. E. (2011), "Endurance time method for multi-component analysis of steel elastic moment frames", Scientia Iranica, v18, n2, p139-142, doi:10.1016/j.scient.2011.03.024
- Earthquake-engineering: aboutcivil.org, Retrieved 2 June, 2019
- Reitherman, Robert (2012). Earthquakes and Engineers: An International History. Reston, VA: ASCE Press. Pp. 356–357. ISBN 9780784410714. Archived from the original on 2012-07-26
- Riahi, H. T. and Estekanchi, H. E, (2010), "Seismic Assessment of Steel Frames with Endurance Time Method", Journal of Constructional Steel Research, v66, n6, p780-792, doi:10.1016/j.jcsr.2009.12.001
- Sloan, S. W. (2013). "Geotechnical stability analysis". Géotechnique. 63 (7): 531. Doi:10.1680/geot.12.RL.001

6

Hydraulic Engineering: A Comprehensive Study

Hydraulic engineering deals with the application of fluid mechanics for solving problems related to the collection, storage, control and use of water. It also makes use of principles from the fields of fluid dynamics and hydrostatics. All these diverse aspects of hydraulic engineering have been carefully analyzed in this chapter.

Hydraulic engineering is a broad field that ranges from the builder to the academic researcher. Without such a range it would not be the dynamic field that it is and, more importantly, it could not have contributed to society in the positive way that it has over the past century, and it would not continue to be a viable, challenging, and important profession. To illustrate the historical perspective to this question, and in so doing illustrate the evolution of hydraulic engineering, the present work uses one of the more visible activities involving hydraulic engineers—large water projects and especially dams in the United States. The reader, though, should not be misled into neglecting the myriad of other activities in which hydraulic engineers engage, some—individually or in combination—equally important to dams. The huge increase over the past 150 years in understanding of flow processes, especially those that occur in nature, and the associated ability to quantify these processes for analysis, design, and prediction is especially important.

However, the direct answer to what is hydraulic engineering does not lie solely in its history. The profession has always been a leader in the use of the latest technology; thus, technological innovation plays a vital part in the modern practice of hydraulic engineering. Innovations include modern computation, including techniques to make detailed flow processes and their complexinteractions with other processes easily understandable. They also include the use of modern electronics for data gathering in the laboratory and the field and a myriad of other tools such as satellite photography, data transmission, global-positioning satellites, geographical data systems, lasers for laboratory and field measurement, radar, lidar, and sonar. Most importantly, they include the hydraulic engineer's interaction with the natural environment and ecology, an interaction that holds great promise and challenge. Indeed, the challenges of the last century, brilliantly solved by the collaboration of academics, small and large private companies, and government action agencies, are being replaced by new demands that will require even more interchange.

Hydraulic engineers have always worked with non-engineering disciplines economists, meteorologists, biologist, and others!, especially in the planning of large projects. Outstanding examples include the salmon problem in the northwest ecological studies along the Missouri River, and the operation of Glen Canyon dam. Such cooperation has steadily increased in the latter part of the twentieth century as its necessity has become more apparent. However, the research community has focused primarily on technical progress rather than on "interdisciplinary studies." Interdisciplinary cooperation is not easy and is frequently frustrating.

First, we have to learn a new language—for example, what is "nature's capital" as opposed to economic capital or "gross natural product" versus gross national product Even more frustrating is the fact that ecological and economic concepts are difficult to express quantitatively, and generally impossible to analyze using the methods of physics. When we bring in social science and humanities, that difficulty is multiplied many times. But a quantitative understanding of such disciplines is no more difficult than explaining the concepts of fluid mechanics, hydraulics, and hydrology to people uninitiated in hydraulic engineering. Finally, researchers who work on such projects are likely to receive little credit. Papers with no mathematics or only simple formula are considered "soft" to fellow engineers, and certainly engineers will have difficulty gaining a foothold in other areas. In research universities, tenure and promotion committees are likely to be suspicious of such interdisciplinary activities. Indeed, such suspicion is well founded and appropriate.

Although we have readily adapted advances in technologically based disciplines—computing, instrumentation, mathematics, numerical methods—into hydraulic engineering, the current challenge is different in that these areas speak the language of engineering and science; they are quantitative. In adapting technology to technological problems, productive and unproductive paths are quickly discovered. In attempting to integrate hydraulic engineering with ecology and social science, it is easy to become bogged down in largely unproductive rhetoric. When the goals are not well defined, the path to those goals cannot be clear.

Hydraulic engineering is the application of fluid mechanics to the liquid earth. Although some applications involve man-made systems pipe networks, for example!, many deal with the complexities of nature. Those latter applications include river engineering; sediment transport; groundwater movement; lake, ocean, and reservoir dynamics including the complications of stratification!; waves; surface flow; and the alteration of natural flows by man, including pollution. Hydraulic engineering is clearly a field for those who love nature and who are comfortable in applying the laws of fluid mechanics for the betterment of mankind while preserving nature. It is a field that has changed in the last half century, but the challenges were never greater than are those of the present day.

The technical ~mechanical Challenges in hydraulic engineering are immense. No one could accurately predict the results of the artificial flood release of Glen Canyon in 1996 toward the objective of partial restoration of nature and natural habitat. Nor is this problem one of traditional hydraulic engineering because "restoration" is not simple; it contains multiple implications in the fields of ecology and biology as well as engineering. The immensity of the problems requires the instrumentation, simulation, and communications tools that increasingly are at our disposal.

In a real sense hydraulic engineering is at a crossroads. The midcentury challenges were met and largely conquered, albeit with inadequate foresight in some cases. That lack of foresight is only a minor factor in the problems of the twenty first century. In the United States, society has the luxury of debate whether to conquer nature for our economic benefit or to preserve nature for the enjoyment of future generations ~and, perhaps, for their economic benefit. Hydraulic engineers can sit on the sidelines and simply do the bidding of the politician or we can influence the debate. Certainly the professional attitude at the very least requires us to lay out the alternatives as we know them and to perform the research to know the alternatives and their consequences as best we can.

The words “water shortage” are often heard. In the United States there is no permanent water shortage, anywhere although temporary shortages may exist. The country is bordered by oceans that have an unlimited supply of water. Water is a commodity and, as such, it can be priced according to its supply and demand. With sufficient engineering works we can supply water anywhere, but of course at a price. Traditional hydraulic engineering is only a part of the determination of that price; the other part is the effect on ecology of the source and of the region that receives the water. Outside of the United States permanent water shortages occur in regions where the resources do not exist for its acquisition, either locally or by importation. Naudascher argues that in such areas public works projects such as big dams have not helped the really needy. However, they have, without doubt, often contributed to the economy and the general welfare of several nations and to their political stability. Neither blanket condemnation nor blanket acceptance of such projects is a reasonable stance. The displacements of peasants by large dams and their inability to benefit from the irrigation, power, and recreation are well known and documented, but such displacements must be balanced against the destruction and forced relocations caused by flooding. According to Naudascher, flood control may contribute to impoverishment in that it eliminates the natural fertilization of the land through the deposition of silt and eliminates the flushing of salt. He also points out that these deprivations may actually be counted as benefits because the fertilizer industry increases sales and irrigation, drainage, and desalinization schemes—necessary due to isolation from the river—are added to the gross national product of the nation. Unfortunately, political power often rests in those that see only one side of the problem. Fortunately, the hydraulic engineering literature contains a much more balanced perspective than is common in the popular media.

Hydrostatics

Fluid statics or hydrostatics is the branch of fluid mechanics that studies “fluids at rest and the pressure in a fluid or exerted by a fluid on an immersed body”.

It encompasses the study of the conditions under which fluids are at rest in stable equilibrium as opposed to fluid dynamics, the study of fluids in motion. Hydrostatics are categorized as a part of the fluid statics, which is the study of all fluids, incompressible or not, at rest.

Hydrostatics is fundamental to hydraulics, the engineering of equipment for storing, transporting and using fluids. It is also relevant to geophysics and astrophysics (for example, in understanding plate tectonics and the anomalies of the Earth’s gravitational field), to meteorology, to medicine (in the context of blood pressure), and many other fields.

Hydrostatics offers physical explanations for many phenomena of everyday life, such as why atmospheric pressure changes with altitude, why wood and oil float on water, and why the surface of still water is always level.

Hydrostatics in Ancient Greece and Rome

Pythagorean Cup

The "fair cup" or Pythagorean cup, which dates from about the 6th century BC, is a hydraulic technology whose invention is credited to the Greek mathematician and geometer Pythagoras. It was used as a learning tool.

The cup consists of a line carved into the interior of the cup, and a small vertical pipe in the center of the cup that leads to the bottom. The height of this pipe is the same as the line carved into the interior of the cup. The cup may be filled to the line without any fluid passing into the pipe in the center of the cup. However, when the amount of fluid exceeds this fill line, fluid will overflow into the pipe in the center of the cup. Due to the drag that molecules exert on one another, the cup will be emptied.

Heron's Fountain

Heron's fountain is a device invented by Heron of Alexandria that consists of a jet of fluid being fed by a reservoir of fluid. The fountain is constructed in such a way that the height of the jet exceeds the height of the fluid in the reservoir, apparently in violation of principles of hydrostatic pressure. The device consisted of an opening and two containers arranged one above the other. The intermediate pot, which was sealed, was filled with fluid, and several cannula (a small tube for transferring fluid between vessels) connecting the various vessels. Trapped air inside the vessels induces a jet of water out of a nozzle, emptying all water from the intermediate reservoir.

Pascal's Contribution in Hydrostatics

Pascal made contributions to developments in both hydrostatics and hydrodynamics. Pascal's Law is a fundamental principle of fluid mechanics that states that any pressure applied to the surface of a fluid is transmitted uniformly throughout the fluid in all directions, in such a way that initial variations in pressure are not changed.

Pressure in Fluids at Rest

Due to the fundamental nature of fluids, a fluid cannot remain at rest under the presence of a shear stress. However, fluids can exert pressure normal to any contacting surface. If a point in the fluid is thought of as an infinitesimally small cube, then it follows from the principles of equilibrium that the pressure on every side of this unit of fluid must be equal. If this were not the case, the fluid would move in the direction of the resulting force. Thus, the pressure on a fluid at rest is isotropic; i.e., it acts with equal magnitude in all directions. This characteristic allows fluids to transmit force through the length of pipes or tubes; i.e., a force applied to a fluid in a pipe is transmitted, via the fluid, to the other end of the pipe. This principle was first formulated, in a slightly extended form, by Blaise Pascal, and is now called Pascal's law.

Hydrostatic Pressure

In a fluid at rest, all frictional and inertial stresses vanish and the state of stress of the system is called *hydrostatic*. When this condition of $V = 0$ is applied to the Navier-Stokes equation, the gradient of pressure becomes a function of body forces only. For a barotropic fluid in a conservative force field like a gravitational force field, pressure exerted by a fluid at equilibrium becomes a function of force exerted by gravity.

The hydrostatic pressure can be determined from a control volume analysis of an infinitesimally small cube of fluid. Since pressure is defined as the force exerted on a test area ($p = F/A$, with p: pressure, F: force normal to area A, A: area), and the only force acting on any such small cube of fluid is the weight of the fluid column above it, hydrostatic pressure can be calculated according to the following formula:

$$p(z)-p(z_0)=\frac{1}{A}\int_{z_0}^{z} dz'\iint_A dx'dy'\rho(z')g(z')=\int_{z_0}^{z} dz'\rho(z')g(z'),$$

where:

- p is the hydrostatic pressure (Pa),
- ρ is the fluid density (kg/m^3),
- g is gravitational acceleration (m/s^2),
- A is the test area (m^2),
- z is the height (parallel to the direction of gravity) of the test area (m),
- z_0 is the height of the zero reference point of the pressure (m).

For water and other liquids, this integral can be simplified significantly for many practical applications, based on the following two assumptions: Since many liquids can be considered incompressible, a reasonable good estimation can be made from assuming a constant density throughout the liquid. (The same assumption cannot be made within a gaseous environment.) Also, since the height h of the fluid column between z and z_0 is often reasonably small compared to the radius of the Earth, one can neglect the variation of g. Under these circumstances, the integral is simplified into the formula:

$$p-p_0=\rho gh,$$

where h is the height $z - z_0$ of the liquid column between the test volume and the zero reference point of the pressure. This formula is often called Stevin's law. Note that this reference point should lie at or below the surface of the liquid. Otherwise, one has to split the integral into two (or more) terms with the constant ρ_{liquid} and $\rho(z')_{\text{above}}$. For example, the absolute pressure compared to vacuum is:

$$p=\rho gH+p_{\text{atm}},$$

where H is the total height of the liquid column above the test area to the surface, and p_{atm} is the atmospheric pressure, i.e., the pressure calculated from the remaining integral over the air column from the liquid surface to infinity. This can easily be visualized using a pressure prism.

Hydrostatic pressure has been used in the preservation of foods in a process called pascalization.

Medicine

In medicine, hydrostatic pressure in blood vessels is the pressure of the blood against the wall. It is the opposing force to oncotic pressure.

Atmospheric Pressure

Statistical mechanics shows that, for a gas of constant temperature, T, its pressure, p will vary with height, h, as:

$$p(h) = p(0)e^{-\frac{Mgh}{kT}}$$

where:

- g is the acceleration due to gravity
- T is the absolute temperature
- k is Boltzmann constant
- M is the mass of a single molecule of gas
- p is the pressure
- h is the height

This is known as the barometric formula, and may be derived from assuming the pressure is hydrostatic.

If there are multiple types of molecules in the gas, the partial pressure of each type will be given by this equation. Under most conditions, the distribution of each species of gas is independent of the other species.

Buoyancy

Any body of arbitrary shape which is immersed, partly or fully, in a fluid will experience the action of a net force in the opposite direction of the local pressure gradient. If this pressure gradient arises from gravity, the net force is in the vertical direction opposite that of the gravitational force. This vertical force is termed buoyancy or buoyant force and is equal in magnitude, but opposite in direction, to the weight of the displaced fluid. Mathematically,

$$F = \rho g V$$

where ρ is the density of the fluid, g is the acceleration due to gravity, and V is the volume of fluid directly above the curved surface. In the case of a ship, for instance, its weight is balanced by pressure forces from the surrounding water, allowing it to float. If more cargo is loaded onto the ship, it would sink more into the water – displacing more water and thus receive a higher buoyant force to balance the increased weight.

Discovery of the principle of buoyancy is attributed to Archimedes.

Hydrostatic Force on Submerged Surfaces

The horizontal and vertical components of the hydrostatic force acting on a submerged surface are given by the following:

$$F_{\mathrm{h}} = p_{\mathrm{c}} A$$

$$F_{\mathrm{v}} = \rho g V$$

where:

- p_{c} is the pressure at the centroid of the vertical projection of the submerged surface
- A is the area of the same vertical projection of the surface
- ρ is the density of the fluid
- g is the acceleration due to gravity
- V is the volume of fluid directly above the curved surface

Liquids (Fluids with Free Surfaces)

Liquids can have free surfaces at which they interface with gases, or with a vacuum. In general, the lack of the ability to sustain a shear stress entails that free surfaces rapidly adjust towards an equilibrium. However, on small length scales, there is an important balancing force from surface tension.

Capillary Action

When liquids are constrained in vessels whose dimensions are small, compared to the relevant length scales, surface tension effects become important leading to the formation of a meniscus through capillary action. This capillary action has profound consequences for biological systems as it is part of one of the two driving mechanisms of the flow of water in plant xylem, the transpirational pull.

Hanging Drops

Without surface tension, drops would not be able to form. The dimensions and stability of drops are determined by surface tension. The drop's surface tension is directly proportional to the cohesion property of the fluid.

Fluid Dynamics

In physics and engineering, fluid dynamics is a subdiscipline of fluid mechanics that describes the flow of fluids—liquids and gases. It has several subdisciplines, including aerodynamics (the study of air and other gases in motion) and hydrodynamics (the study of liquids in motion). Fluid dynamics has a wide range of applications, including calculating forces and moments on aircraft, determining the mass flow rate of petroleum through pipelines, predicting weather patterns, understanding nebulae in interstellar space and modelling fission weapon detonation,

Fluid dynamics offers a systematic structure—which underlies these practical disciplines—that embraces empirical and semi-empirical laws derived from flow measurement and used to solve practical problems. The solution to a fluid dynamics problem typically involves the calculation of various properties of the fluid, such as flow velocity, pressure, density, and temperature, as functions of space and time.

Before the twentieth century, *hydrodynamics* was synonymous with fluid dynamics. This is still reflected in names of some fluid dynamics topics, like magnetohydrodynamics and hydrodynamic stability, both of which can also be applied to gases.

The foundational axioms of fluid dynamics are the conservation laws, specifically, conservation of mass, conservation of linear momentum, and conservation of energy (also known as First Law of Thermodynamics). These are based on classical mechanics and are modified in quantum mechanics and general relativity. They are expressed using the Reynolds transport theorem.

In addition to the above, fluids are assumed to obey the continuum assumption. Fluids are composed of molecules that collide with one another and solid objects. However, the continuum assumption assumes that fluids are continuous, rather than discrete. Consequently, it is assumed that properties such as density, pressure, temperature, and flow velocity are well-defined at infinitesimally small points in space and vary continuously from one point to another. The fact that the fluid is made up of discrete molecules is ignored.

For fluids that are sufficiently dense to be a continuum, do not contain ionized species, and have flow velocities small in relation to the speed of light, the momentum equations for Newtonian fluids are the Navier–Stokes equations—which is a non-linear set of differential equations that describes the flow of a fluid whose stress depends linearly on flow velocity gradients and pressure. The unsimplified equations do not have a general closed-form solution, so they are primarily of use in Computational Fluid Dynamics. The equations can be simplified in a number of ways, all of which make them easier to solve. Some of the simplifications allow some simple fluid dynamics problems to be solved in closed form.

In addition to the mass, momentum, and energy conservation equations, a thermodynamic equation of state that gives the pressure as a function of other thermodynamic variables is required to completely describe the problem. An example of this would be the perfect gas equation of state:

$$p = \frac{\rho R_u T}{M}$$

where p is pressure, ρ is density, T the absolute temperature, while R_u is the gas constant and M is molar mass for a particular gas.

Conservation Laws

Three conservation laws are used to solve fluid dynamics problems, and may be written in integral or differential form. The conservation laws may be applied to a region of the flow called a *control volume*. A control volume is a discrete volume in space through which fluid is assumed to flow. The integral formulations of the conservation laws are used to describe the change of mass, momentum, or energy within the control volume. Differential formulations of the conservation laws apply Stokes' theorem to yield an expression which may be interpreted as the integral form of the law applied to an infinitesimally small volume (at a point) within the flow.

Mass continuity (conservation of mass): The rate of change of fluid mass inside a control volume must be equal to the net rate of fluid flow into the volume. Physically, this statement requires that mass is neither created nor destroyed in the control volume, and can be translated into the integral form of the continuity equation:

$$\frac{\partial}{\partial t}\iiint_V \rho \, dV = -\oiint_S \rho \mathbf{u} \cdot d\mathbf{S}$$

Above, ρ is the fluid density, u is the flow velocity vector, and t is time. The left-hand side of the above expression is the rate of increase of mass within the volume and contains a triple integral over the control volume, whereas the right-hand side contains an integration over the surface of the control volume of mass convected into the system. Mass flow into the system is accounted as positive, and since the normal vector to the surface is opposite the sense of flow into the system the term is negated. The differential form of the continuity equation is, by the divergence theorem:

$$\frac{\partial \rho}{\partial t} + \nabla \cdot (\rho \mathbf{u}) = 0$$

Conservation of momentum: Newton's second law of motion applied to a control volume, is a statement that any change in momentum of the fluid within that control volume will be due to the net flow of momentum into the volume and the action of external forces acting on the fluid within the volume.

$$\frac{\partial}{\partial t}\iiint_V \rho \mathbf{u} \, dV = -\oiint_S (\rho \mathbf{u}.dS)\mathbf{u} - \oiint_S p\, dS + \iiint_V \rho \mathbf{f}_{\text{body}} \, dV + \mathbf{F}_{\text{surf}}$$

In the above integral formulation of this equation, the term on the left is the net change of momentum within the volume. The first term on the right is the net rate at which momentum is convected into the volume. The second term on the right is the force due to pressure on the volume's surfaces. The first two terms on the right are negated since momentum entering the system is accounted as positive, and the normal is opposite the direction of the velocity $\mathbf{u}$ and pressure forces. The third term on the right is the net acceleration of the mass within the volume due to any body forces (here represented by f_{body}). Surface forces, such as viscous forces, are represented by $\mathbf{F}_{\text{surf}}$, the net force due to shear forces acting on the volume surface. The momentum balance can also be written for a *moving* control volume.

The following is the differential form of the momentum conservation equation. Here, the volume is reduced to an infinitesimally small point, and both surface and body forces are accounted for in one total force, F. For example, F may be expanded into an expression for the frictional and gravitational forces acting at a point in a flow.

$$\frac{D\mathbf{u}}{Dt} = \mathbf{F} - \frac{\nabla p}{\rho}$$

In aerodynamics, air is assumed to be a Newtonian fluid, which posits a linear relationship between the shear stress (due to internal friction forces) and the rate of strain of the fluid. The equation above is a vector equation in a three-dimensional flow, but it can be expressed as three scalar equations in three coordinate directions. The conservation of momentum equations for the compressible, viscous flow case are called the Navier–Stokes equations.

Conservation of energy: Although energy can be converted from one form to another, the total energy in a closed system remains constant,

$$\rho \frac{Dh}{Dt} = \frac{Dp}{Dt} + \nabla \cdot \left(k \nabla T \right) + \Phi$$

Above, h is enthalpy, k is the thermal conductivity of the fluid, T is temperature, and Ö is the viscous dissipation function. The viscous dissipation function governs the rate at which mechanical energy of the flow is converted to heat. The second law of thermodynamics requires that the dissipation term is always positive: viscosity cannot create energy within the control volume. The expression on the left side is a material derivative.

Compressible vs. Incompressible Flow

All fluids are compressible to some extent; that is, changes in pressure or temperature cause changes in density. However, in many situations the changes in pressure and temperature are sufficiently small that the changes in density are negligible. In this case the flow can be modelled as an incompressible flow. Otherwise the more general compressible flow equations must be used.

Mathematically, incompressibility is expressed by saying that the density ρ of a fluid parcel does not change as it moves in the flow field, i.e.,

$$\frac{\mathrm{D}\rho}{\mathrm{D}t} = 0,$$

where D/Dt is the material derivative, which is the sum of local and convective derivatives. This additional constraint simplifies the governing equations, especially in the case when the fluid has a uniform density.

For flow of gases, to determine whether to use compressible or incompressible fluid dynamics, the Mach number of the flow is evaluated. As a rough guide, compressible effects can be ignored at Mach numbers below approximately 0.3. For liquids, whether the incompressible assumption is valid depends on the fluid properties (specifically the critical pressure and temperature of the fluid) and the flow conditions (how close to the critical pressure the actual flow pressure becomes). Acoustic problems always require allowing compressibility, since sound waves are compression waves involving changes in pressure and density of the medium through which they propagate.

Newtonian vs non-Newtonian Fluids

Flow around an airfoil.

All fluids are viscous, meaning that they exert some resistance to deformation: neighbouring parcels of fluid moving at different velocities exert viscous forces on each other. The velocity gradient is referred to as a strain rate; it has dimensions T^{-1}. Isaac Newton showed that for many familiar fluids such as water and air, the stress due to these viscous forces is linearly related to the strain rate. Such fluids are called Newtonian fluids. The coefficient of proportionality is called the fluid's viscosity; for Newtonian fluids, it is a fluid property that is independent of the strain rate.

Non-Newtonian fluids have a more complicated, non-linear stress-strain behaviour. The sub-discipline of rheology describes the stress-strain behaviours of such fluids, which include emulsions and slurries, some viscoelastic materials such as blood and some polymers, and *sticky liquids* such as latex, honey and lubricants.

Inviscid vs. Viscous vs. Stokes Flow

The dynamic of fluid parcels is described with the help of Newton's second law. An accelerating parcel of fluid is subject to inertial effects.

The Reynolds number is a dimensionless quantity which characterises the magnitude of inertial effects compared to the magnitude of viscous effects. A low Reynolds number ($Re \ll 1$) indicates that viscous forces are very strong compared to inertial forces. In such cases, inertial forces are sometimes neglected; this flow regime is called Stokes or creeping flow.

In contrast, high Reynolds numbers ($Re \gg 1$) indicate that the inertial effects have more effect on the velocity field than the viscous (friction) effects. In high Reynolds number flows, the flow is often modeled as an inviscid flow, an approximation in which viscosity is completely neglected. Eliminating viscosity allows the Navier–Stokes equations to be simplified into the Euler equations. The integration of the Euler equations along a streamline in an inviscid flow yields Bernoulli's equation. When, in addition to being inviscid, the flow is irrotational everywhere, Bernoulli's equation can completely describe the flow everywhere. Such flows are called potential flows, because the velocity field may be expressed as the gradient of a potential energy expression.

This idea can work fairly well when the Reynolds number is high. However, problems such as those involving solid boundaries may require that the viscosity be included. Viscosity cannot be neglected near solid boundaries because the no-slip condition generates a thin region of large strain rate, the boundary layer, in which viscosity effects dominate and which thus generates vorticity. Therefore, to calculate net forces on bodies (such as wings), viscous flow equations must be used: inviscid flow theory fails to predict drag forces, a limitation known as the d'Alembert's paradox.

A commonly used model, especially in computational fluid dynamics, is to use two flow models: the Euler equations away from the body, and boundary layer equations in a region close to the body. The two solutions can then be matched with each other, using the method of matched asymptotic expansions.

Steady vs. Unsteady Flow

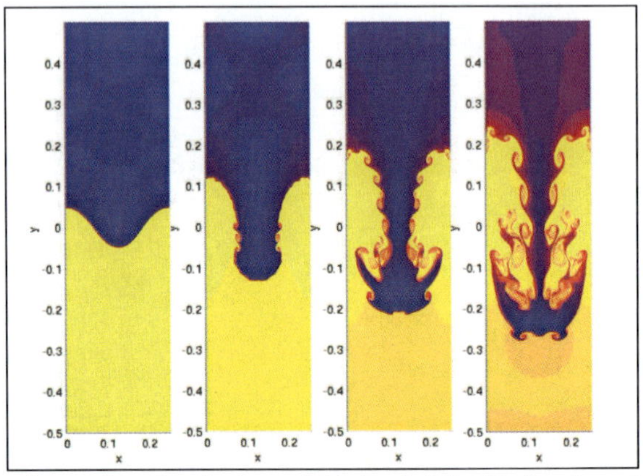

Hydrodynamics simulation of the Rayleigh–Taylor instability.

A flow that is not a function of time is called steady flow. Steady-state flow refers to the condition where the fluid properties at a point in the system do not change over time. Time dependent flow is known as unsteady (also called transient). Whether a particular flow is steady or unsteady, can depend on the chosen frame of reference. For instance, laminar flow over a sphere is steady in the frame of reference that is stationary with respect to the sphere. In a frame of reference that is stationary with respect to a background flow, the flow is unsteady.

A turbulent flow can, however, be statistically stationary. According to Pope:

- The random field $U(x,t)$ is statistically stationary, if all statistics are invariant under a shift in time.
- This roughly means that all statistical properties are constant in time. Often, the mean field is the object of interest, and this is constant too in a statistically stationary flow.
- Steady flows are often more tractable than otherwise similar unsteady flows. The governing equations of a steady problem have one dimension fewer (time) than the governing equations of the same problem without taking advantage of the steadiness of the flow field.

Laminar vs. Turbulent Flow

Turbulence is flow characterized by recirculation, eddies, and apparent randomness. Flow in which turbulence is not exhibited is called laminar. The presence of eddies or recirculation alone does not necessarily indicate turbulent flow—these phenomena may be present in laminar flow as well. Mathematically, turbulent flow is often represented via a Reynolds decomposition, in which the flow is broken down into the sum of an average component and a perturbation component.

It is believed that turbulent flows can be described well through the use of the Navier–Stokes equations. Direct numerical simulation (DNS), based on the Navier–Stokes equations, makes it possible to simulate turbulent flows at moderate Reynolds numbers. Restrictions depend on the power of the computer used and the efficiency of the solution algorithm. The results of DNS have been found to agree well with experimental data for some flows.

Most flows of interest have Reynolds numbers much too high for DNS to be a viable option, given the state of computational power for the next few decades. Any flight vehicle large enough to carry a human (L > 3 m), moving faster than 20 m/s (72 km/h) is well beyond the limit of DNS simulation (Re = 4 million). Transport aircraft wings (such as on an Airbus A300 or Boeing 747) have Reynolds numbers of 40 million (based on the wing chord dimension). Solving these real-life flow problems requires turbulence models for the foreseeable future. Reynolds-averaged Navier–Stokes equations (RANS) combined with turbulence modelling provides a model of the effects of the turbulent flow. Such a modelling mainly provides the additional momentum transfer by the Reynolds stresses, although the turbulence also enhances the heat and mass transfer. Another promising methodology is large eddy simulation (LES), especially in the guise of detached eddy simulation (DES)—which is a combination of RANS turbulence modelling and large eddy simulation.

Subsonic vs. Transonic, Supersonic and Hypersonic Flows

While many flows (e.g. flow of water through a pipe) occur at low Mach numbers, many flows of practical interest in aerodynamics or in turbomachines occur at high fractions of M=1 (transonic flows) or in excess of it (supersonic or even hypersonic flows). New phenomena occur at these regimes such as instabilities in transonic flow, shock waves for supersonic flow, or non-equilibrium chemical behaviour due to ionization in hypersonic flows. In practice, each of those flow regimes is treated separately.

Reactive vs. Non-Reactive Flows

Reactive flows are flows that are chemically reactive, which finds its applications in many areas such as combustion(IC engine), propulsion devices (Rockets, jet engines etc.), detonations, fire and safety hazards, astrophysics etc. In addition to conservation of mass, momentum and energy, conservation of individual species (for example, mass fraction of methane in methane combustion) need to be derived, where the production/depletion rate of any species are obtained by simultaneously solving the equations of chemical kinetics.

Magnetohydrodynamics

Magnetohydrodynamics is the multi-disciplinary study of the flow of electrically conducting fluids in electromagnetic fields. Examples of such fluids include plasmas, liquid metals, and salt water. The fluid flow equations are solved simultaneously with Maxwell's equations of electromagnetism.

Relativistic Fluid Dynamics

Relativistic fluid dynamics studies the macroscopic and microscopic fluid motion at large velocities comparable to the velocity of light. This branch of fluid dynamics accounts the relativistic effects both from the special theory of relativity and the general theory of relativity. The governing equations are derived in Riemannian geometry for Minkowski spacetime.

Other Approximations

There are a large number of other possible approximations to fluid dynamic problems. Some of the more commonly used are listed below.

- The Boussinesq approximation neglects variations in density except to calculate buoyancy forces. It is often used in free convection problems where density changes are small.
- Lubrication theory and Hele–Shaw flow exploits the large aspect ratio of the domain to show that certain terms in the equations are small and so can be neglected.
- Slender-body theory is a methodology used in Stokes flow problems to estimate the force on, or flow field around, a long slender object in a viscous fluid.
- The shallow-water equations can be used to describe a layer of relatively inviscid fluid with a free surface, in which surface gradients are small.
- Darcy's law is used for flow in porous media, and works with variables averaged over several pore-widths.
- In rotating systems, the quasi-geostrophic equations assume an almost perfect balance between pressure gradients and the Coriolis force. It is useful in the study of atmospheric dynamics.

Bernoulli's Principle

In fluid dynamics, Bernoulli's principle states that an increase in the speed of a fluid occurs simultaneously with a decrease in pressure or a decrease in the fluid's potential energy. The principle is named after Daniel Bernoulli who published it in his book Hydrodynamica in 1738. Although Bernoulli deduced that pressure decreases when the flow speed increases, it was Leonhard Euler who derived Bernoulli's equation in its usual form in 1752. The principle is only applicable for isentropic flows: when the effects of irreversible processes (like turbulence) and non-adiabatic processes (e.g. heat radiation) are small and can be neglected.

Bernoulli's principle can be applied to various types of fluid flow, resulting in various forms of *Bernoulli's equation*; there are different forms of Bernoulli's equation for different types of flow. The simple form of Bernoulli's equation is valid for incompressible flows (e.g. most liquid flows and gases moving at low Mach number). More advanced forms may be applied to compressible flows at higher Mach numbers.

Bernoulli's principle can be derived from the principle of conservation of energy. This states that, in a steady flow, the sum of all forms of energy in a fluid along a streamline is the same at all points on that streamline. This requires that the sum of kinetic energy, potential energy and internal energy remains constant. Thus an increase in the speed of the fluid – implying an increase in its kinetic energy (dynamic pressure) – occurs with a simultaneous decrease in (the sum of) its potential energy (including the static pressure) and internal energy. If the fluid is flowing out of a reservoir, the sum of all forms of energy is the same on all streamlines because in a reservoir the energy per unit volume (the sum of pressure and gravitational potential $\rho g h$) is the same everywhere.

Bernoulli's principle can also be derived directly from Isaac Newton's Second Law of Motion. If a small volume of fluid is flowing horizontally from a region of high pressure to a region of low pressure, then there is more pressure behind than in front. This gives a net force on the volume, accelerating it along the streamline.

Fluid particles are subject only to pressure and their own weight. If a fluid is flowing horizontally and along a section of a streamline, where the speed increases it can only be because the fluid on that section has moved from a region of higher pressure to a region of lower pressure; and if its speed decreases, it can only be because it has moved from a region of lower pressure to a region of higher pressure. Consequently, within a fluid flowing horizontally, the highest speed occurs where the pressure is lowest, and the lowest speed occurs where the pressure is highest.

Incompressible Flow Equation

In most flows of liquids, and of gases at low Mach number, the density of a fluid parcel can be considered to be constant, regardless of pressure variations in the flow. Therefore, the fluid can be considered to be incompressible and these flows are called incompressible flows. Bernoulli performed his experiments on liquids, so his equation in its original form is valid only for incompressible flow. A common form of Bernoulli's equation, valid at any arbitrary point along a streamline, is:

$$\frac{v^2}{2} + gz + \frac{p}{\rho} = \text{constant}$$

where:

- v is the fluid flow speed at a point on a streamline,
- g is the acceleration due to gravity,
- z is the elevation of the point above a reference plane, with the positive z-direction pointing upward – so in the direction opposite to the gravitational acceleration,
- p is the pressure at the chosen point, and
- ρ is the density of the fluid at all points in the fluid.

The constant on the right-hand side of the equation depends only on the streamline chosen, whereas v, z and p depend on the particular point on that streamline.

The following assumptions must be met for this Bernoulli equation to apply:

- The flow must be steady, i.e. the flow parameters (velocity, density, etc.) at any point cannot change with time,
- The flow must be incompressible – even though pressure varies, the density must remain constant along a streamline;
- Friction by viscous forces must be negligible.

For conservative force fields (not limited to the gravitational field), Bernoulli's equation can be generalized as:

$$\frac{v^2}{2} + \Psi + \frac{p}{\rho} = \text{constant}$$

where Ψ is the force potential at the point considered on the streamline. E.g. for the Earth's gravity $\Psi = gz$.

By multiplying with the fluid density ρ, equation (**A**) can be rewritten as:

$$\tfrac{1}{2}\rho v^2 + \rho g z + p = \text{constant}$$

or

$$q + \rho g h = p_0 + \rho g z = \text{constant}$$

where,

- $q = 1/2\rho v^2$ is dynamic pressure,
- $h = z + p/\rho g$ is the piezometric head or hydraulic head (the sum of the elevation z and the pressure head) and
- $p_o = p + q$ is the total pressure (the sum of the static pressure p and dynamic pressure q).

The constant in the Bernoulli equation can be normalised. A common approach is in terms of total head or energy head H:

$$H = z + \frac{p}{\rho g} + \frac{v^2}{2g} = h + \frac{v^2}{2g},$$

The above equations suggest there is a flow speed at which pressure is zero, and at even higher speeds the pressure is negative. Most often, gases and liquids are not capable of negative absolute pressure, or even zero pressure, so clearly Bernoulli's equation ceases to be valid before zero pressure is reached. In liquids – when the pressure becomes too low – cavitation occurs. The above equations use a linear relationship between flow speed squared and pressure. At higher flow speeds in gases, or for sound waves in liquid, the changes in mass density become significant so that the assumption of constant density is invalid.

Simplified Form

In many applications of Bernoulli's equation, the change in the $\rho g z$ term along the streamline is so small compared with the other terms that it can be ignored. For example, in the case of aircraft in flight, the change in height z along a streamline is so small the $\rho g z$ term can be omitted. This allows the above equation to be presented in the following simplified form:

$$p + q = p_0$$

where p_o is called "total pressure", and q is "dynamic pressure". Many authors refer to the pressure p as static pressure to distinguish it from total pressure p_0 and dynamic pressure q. In *Aerodynamics*, L.J. Clancy writes: "To distinguish it from the total and dynamic pressures, the actual pressure of the fluid, which is associated not with its motion but with its state, is often referred to as the static pressure, but where the term pressure alone is used it refers to this static pressure."

The simplified form of Bernoulli's equation can be summarized in the following memorable word equation:

static pressure + dynamic pressure = total pressure

Every point in a steadily flowing fluid, regardless of the fluid speed at that point, has its own unique static pressure p and dynamic pressure q. Their sum $p + q$ is defined to be the total pressure p_0. The significance of Bernoulli's principle can now be summarized as "total pressure is constant along a streamline".

If the fluid flow is irrotational, the total pressure on every streamline is the same and Bernoulli's principle can be summarized as "total pressure is constant everywhere in the fluid flow". It is reasonable to assume that irrotational flow exists in any situation where a large body of fluid is flowing past a solid body. Examples are aircraft in flight, and ships moving in open bodies of water. However, it is important to remember that Bernoulli's principle does not apply in the boundary layer or in fluid flow through long pipes.

If the fluid flow at some point along a streamline is brought to rest, this point is called a stagnation point, and at this point the total pressure is equal to the stagnation pressure.

Applicability of Incompressible Flow Equation to Flow of Gases

Bernoulli's equation is sometimes valid for the flow of gases: provided that there is no transfer of kinetic or potential energy from the gas flow to the compression or expansion of the gas. If both the gas pressure and volume change simultaneously, then work will be done on or by the gas. In this case, Bernoulli's equation – in its incompressible flow form – cannot be assumed to be valid. However, if the gas process is entirely isobaric, or isochoric, then no work is done on or by the gas, (so the simple energy balance is not upset). According to the gas law, an isobaric or isochoric process is ordinarily the only way to ensure constant density in a gas. Also the gas density will be proportional to the ratio of pressure and absolute temperature, however this ratio will vary upon compression or expansion, no matter what non-zero quantity of heat is added or removed. The only exception is if the net heat transfer is zero, as in a complete thermodynamic cycle, or in an individual isentropic (frictionless adiabatic) process, and even then this reversible process must be reversed, to restore the gas to the original pressure and specific volume, and thus density. Only then is the original, unmodified Bernoulli equation applicable. In this case the equation can be used if the flow speed of the gas is sufficiently below the speed of sound, such that the variation in density of the gas (due to this effect) along each streamline can be ignored. Adiabatic flow at less than Mach 0.3 is generally considered to be slow enough.

Unsteady Potential Flow

The Bernoulli equation for unsteady potential flow is used in the theory of ocean surface waves and acoustics.

For an irrotational flow, the flow velocity can be described as the gradient $\nabla\varphi$ of a velocity potential φ. In that case, and for a constant density ρ, the momentum equations of the Euler equations can be integrated to:

$$\frac{\partial\varphi}{\partial t}+\tfrac{1}{2}v^2+\frac{p}{\rho}+gz=f(t),$$

which is a Bernoulli equation valid also for unsteady—or time dependent—flows. Here $\partial\varphi/\partial t$ denotes the partial derivative of the velocity potential φ with respect to time t, and $v = |\nabla\varphi|$ is the flow speed. The function $f(t)$ depends only on time and not on position in the fluid. As a result, the Bernoulli equation at some moment t does not only apply along a certain streamline, but in the whole fluid domain. This is also true for the special case of a steady irrotational flow, in which case f and $\partial\varphi/\partial t$ are constants so equation (A) can be applied in every point of the fluid domain.

Further $f(t)$ can be made equal to zero by incorporating it into the velocity potential using the transformation

$$\Phi=\varphi-\int_{t_0}^{t} f(\tau)\,\mathrm{d}\tau,$$

resulting in

$$\frac{\partial\Phi}{\partial t}+\tfrac{1}{2}v^2+\frac{p}{\rho}+gz=0.$$

Note that the relation of the potential to the flow velocity is unaffected by this transformation: $\nabla\Phi = \nabla\varphi$.

The Bernoulli equation for unsteady potential flow also appears to play a central role in Luke's variational principle, a variational description of free-surface flows using the Lagrangian.

Compressible Flow Equation

Bernoulli developed his principle from his observations on liquids, and his equation is applicable only to incompressible fluids, and steady compressible fluids up to approximately Mach number 0.3. It is possible to use the fundamental principles of physics to develop similar equations applicable to compressible fluids. There are numerous equations, each tailored for a particular application, but all are analogous to Bernoulli's equation and all rely on nothing more than the fundamental principles of physics such as Newton's laws of motion or the first law of thermodynamics.

Compressible Flow in Fluid Dynamics

For a compressible fluid, with a barotropic equation of state, and under the action of conservative forces,

$$\frac{v^2}{2}+\int_{p_1}^{p}\frac{\mathrm{d}\tilde{p}}{\rho(\tilde{p})}+\Psi=\text{constant (along a streamline)}$$

where:

- p is the pressure.
- ρ is the density and $\rho(\tilde{p})$ indicates that it is a function of pressure.
- v is the flow speed.
- Ψ is the potential associated with the conservative force field, often the gravitational potential.

In engineering situations, elevations are generally small compared to the size of the Earth, and the time scales of fluid flow are small enough to consider the equation of state as adiabatic. In this case, the above equation for an ideal gas becomes:

$$\frac{v^2}{2} + gz + \left(\frac{\gamma}{\gamma - 1}\right)\frac{p}{\rho} = \text{constant (along a streamline)}$$

where, in addition to the terms listed above:

- γ is the ratio of the specific heats of the fluid.
- g is the acceleration due to gravity.
- z is the elevation of the point above a reference plane.

In many applications of compressible flow, changes in elevation are negligible compared to the other terms, so the term gz can be omitted. A very useful form of the equation is then:

$$\frac{v^2}{2} + \left(\frac{\gamma}{\gamma - 1}\right)\frac{p}{\rho} = \left(\frac{\gamma}{\gamma - 1}\right)\frac{p_0}{\rho_0}$$

where,

- p_0 is the total pressure.
- ρ_0 is the total density.

Compressible Flow in Thermodynamics

The most general form of the equation, suitable for use in thermodynamics in case of (quasi) steady flow, is:

$$\frac{v^2}{2} + \Psi + w = \text{constant.}$$

Here w is the enthalpy per unit mass (also known as specific enthalpy), which is also often written as h.

Note that $w = \varepsilon + p/\rho$ where ε is the thermodynamic energy per unit mass, also known as the specific internal energy. So, for constant internal energy ε the equation reduces to the incompressible-flow form.

The constant on the right-hand side is often called the Bernoulli constant, and denoted *b*. For steady inviscid adiabatic flow with no additional sources or sinks of energy, *b* is constant along any given streamline. More generally, when *b* may vary along streamlines, it still proves a useful parameter, related to the "head" of the fluid.

When the change in Ψ can be ignored, a very useful form of this equation is:

$$\frac{v^2}{2} + w = w_0$$

where w_0 is total enthalpy. For a calorically perfect gas such as an ideal gas, the enthalpy is directly proportional to the temperature, and this leads to the concept of the total (or stagnation) temperature.

When shock waves are present, in a reference frame in which the shock is stationary and the flow is steady, many of the parameters in the Bernoulli equation suffer abrupt changes in passing through the shock. The Bernoulli parameter itself, however, remains unaffected. An exception to this rule is radiative shocks, which violate the assumptions leading to the Bernoulli equation, namely the lack of additional sinks or sources of energy.

Derivations of the Bernoulli Equation

Bernoulli Equation for Incompressible Fluids

The Bernoulli equation for incompressible fluids can be derived by either integrating Newton's second law of motion or by applying the law of conservation of energy between two sections along a streamline, ignoring viscosity, compressibility, and thermal effects.

Derivation through Integrating Newton's Second Law of Motion

The simplest derivation is to first ignore gravity and consider constrictions and expansions in pipes that are otherwise straight, as seen in Venturi effect. Let the *x* axis be directed down the axis of the pipe.

Define a parcel of fluid moving through a pipe with cross-sectional area *A*, the length of the parcel is d*x*, and the volume of the parcel *A* d*x*. If mass density is ρ, the mass of the parcel is density multiplied by its volume $m = \rho A\,\mathrm{d}x$. The change in pressure over distance d*x* is d*p* and flow velocity $v = \frac{dx}{dt}$.

Apply Newton's second law of motion (force = mass × acceleration) and recognizing that the effective force on the parcel of fluid is $-A\,\mathrm{d}p$. If the pressure decreases along the length of the pipe, d*p* is negative but the force resulting in flow is positive along the *x* axis.

$$m\frac{\mathrm{d}v}{\mathrm{d}t} = F$$

$$\rho A \mathrm{d}x \frac{\mathrm{d}v}{\mathrm{d}t} = -A\mathrm{d}p$$

$$\rho\frac{\mathrm{d}v}{\mathrm{d}t} = -\frac{\mathrm{d}p}{\mathrm{d}x}$$

In steady flow the velocity field is constant with respect to time, $v = v(x) = v(x(t))$, so v itself is not directly a function of time t. It is only when the parcel moves through x that the cross sectional area changes: v depends on t only through the cross-sectional position $x(t)$.

$$\frac{dv}{dt} = \frac{dv}{dx}\frac{dx}{dt}$$
$$= \frac{dv}{dx}v$$
$$= \frac{d}{dx}\left(\frac{v^2}{2}\right).$$

With density ρ constant, the equation of motion can be written as,

$$\frac{d}{dx}\left(\rho\frac{v^2}{2} + p\right) = 0$$

by integrating with respect to x,

$$\frac{v^2}{2} + \frac{p}{\rho} = C$$

where C is a constant, sometimes referred to as the Bernoulli constant. It is not a universal constant, but rather a constant of a particular fluid system. The deduction is: where the speed is large, pressure is low and vice versa.

In the above derivation, no external work–energy principle is invoked. Rather, Bernoulli's principle was derived by a simple manipulation of Newton's second law.

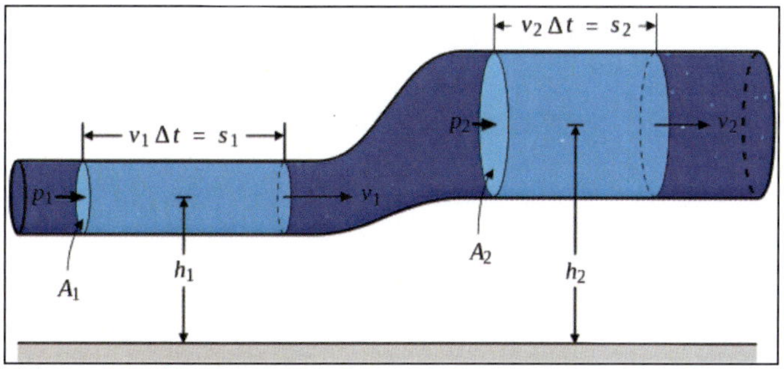

A streamtube of fluid moving to the right. Indicated are pressure, elevation, flow speed, distance (s), and cross-sectional area. Note that in this figure elevation is denoted as h, contrary to the text where it is given by z.

Derivation by using Conservation of Energy

Another way to derive Bernoulli's principle for an incompressible flow is by applying conservation of energy. In the form of the work-energy theorem, stating that, the change in the kinetic energy E_{kin} of the system equals the net work W done on the system;

$$W = \Delta E_{kin}.$$

Therefore,

The work done by the forces in the fluid equals increase in kinetic energy.

The system consists of the volume of fluid, initially between the cross-sections A_1 and A_2. In the time interval Δt fluid elements initially at the inflow cross-section A_1 move over a distance $s_1 = v_1 \Delta t$, while at the outflow cross-section the fluid moves away from cross-section A_2 over a distance $s_2 = v_2 \Delta t$. The displaced fluid volumes at the inflow and outflow are respectively $A_1 s_1$ and $A_2 s_2$. The associated displaced fluid masses are – when ρ is the fluid's mass density – equal to density times volume, so $\rho A_1 s_1$ and $\rho A_2 s_2$. By mass conservation, these two masses displaced in the time interval Δt have to be equal, and this displaced mass is denoted by Δm:

$$\rho A_1 s_1 = \rho A_1 v_1 \Delta t = \Delta m,$$
$$\rho A_2 s_2 = \rho A_2 v_2 \Delta t = \Delta m.$$

The work done by the forces consists of two parts:

- The *work done by the pressure* acting on the areas A_1 and A_2:

$$W_{\text{pressure}} = F_{1,\text{pressure}} s_1 - F_{2,\text{pressure}} s_2 = p_1 A_1 s_1 - p_2 A_2 s_2 = \Delta m \frac{p_1}{\rho} - \Delta m \frac{p_2}{\rho}.$$

- The *work done by gravity*: the gravitational potential energy in the volume $A_1 s_1$ is lost, and at the outflow in the volume $A_2 s_2$ is gained. So, the change in gravitational potential energy $\Delta E_{\text{pot,gravity}}$ in the time interval Δt is:

$$\Delta E_{\text{pot,gravity}} = \Delta m g z_2 - \Delta m g z_1.$$

Now, the work by the force of gravity is opposite to the change in potential energy, $W_{\text{gravity}} = -\Delta E_{\text{pot,gravity}}$: while the force of gravity is in the negative z-direction, the work—gravity force times change in elevation—will be negative for a positive elevation change $\Delta z = z_2 - z_1$, while the corresponding potential energy change is positive. So:

$$W_{\text{gravity}} = -\Delta E_{\text{pot,gravity}} = \Delta m g z_1 - \Delta m g z_2.$$

And therefore the total work done in this time interval Δt is

$$W = W_{\text{pressure}} + W_{\text{gravity}}.$$

The *increase in kinetic energy* is

$$\Delta E_{\text{kin}} = \tfrac{1}{2} \Delta m v_2^2 - \tfrac{1}{2} \Delta m v_1^2.$$

Putting these together, the work-kinetic energy theorem $W = \Delta E_{\text{kin}}$ gives:

$$\Delta m \frac{p_1}{\rho} - \Delta m \frac{p_2}{\rho} + \Delta m g z_1 - \Delta m g z_2 = \tfrac{1}{2} \Delta m v_2^2 - \tfrac{1}{2} \Delta m v_1^2$$

or

$$\tfrac{1}{2}\Delta m v_1^2 + \Delta m g z_1 + \Delta m \frac{p_1}{\rho} = \tfrac{1}{2}\Delta m v_2^2 + \Delta m g z_2 + \Delta m \frac{p_2}{\rho}.$$

After dividing by the mass $\Delta m = \rho A_1 v_1 \Delta t = \rho A_2 v_2 \Delta t$ the result is:

$$\tfrac{1}{2}v_1^2 + g z_1 + \frac{p_1}{\rho} = \tfrac{1}{2}v_2^2 + g z_2 + \frac{p_2}{\rho}$$

or, as stated in the first paragraph:

$$\frac{v^2}{2} + gz + \frac{p}{\rho} = C$$

Further division by g produces the following equation. Note that each term can be described in the length dimension (such as meters). This is the head equation derived from Bernoulli's principle:

$$\frac{v^2}{2g} + z + \frac{p}{\rho g} = C$$

The middle term, z, represents the potential energy of the fluid due to its elevation with respect to a reference plane. Now, z is called the elevation head and given the designation $z_{\text{elevation}}$.

A free falling mass from an elevation $z > 0$ (in a vacuum) will reach a speed

$$v = \sqrt{2gz},$$

when arriving at elevation $z = 0$. Or when we rearrange it as a *head*:

$$h_v = \frac{v^2}{2g}$$

The term $\frac{v^2}{2g}$ is called the *velocity head*, expressed as a length measurement. It represents the internal energy of the fluid due to its motion.

The hydrostatic pressure p is defined as

$$p = p_0 - \rho g z,$$

with p_0 some reference pressure, or when we rearrange it as a *head*:

$$\psi = \frac{p}{\rho g}.$$

The term $\frac{p}{\rho g}$ is also called the *pressure head*, expressed as a length measurement. It represents the internal energy of the fluid due to the pressure exerted on the container. When we combine the head due to the flow speed and the head due to static pressure with the elevation above a reference plane, we obtain a simple relationship useful for incompressible fluids using the velocity head, elevation head, and pressure head.

$$\psi = h_v + z_{\text{elevation}} + \psi = C$$

If we were to multiply Eqn. 1 by the density of the fluid, we would get an equation with three pressure terms:

$$\frac{\rho v^2}{2} + \rho g z + p = C$$

We note that the pressure of the system is constant in this form of the Bernoulli equation. If the static pressure of the system (the far right term) increases, and if the pressure due to elevation (the middle term) is constant, then we know that the dynamic pressure (the left term) must have decreased. In other words, if the speed of a fluid decreases and it is not due to an elevation difference, we know it must be due to an increase in the static pressure that is resisting the flow.

All three equations are merely simplified versions of an energy balance on a system.

Bernoulli Equation for Compressible Fluids

The derivation for compressible fluids is similar. Again, the derivation depends upon (1) conservation of mass, and (2) conservation of energy. Conservation of mass implies that in the above figure, in the interval of time Δt, the amount of mass passing through the boundary defined by the area A_1 is equal to the amount of mass passing outwards through the boundary defined by the area A_2:

$$0 = \Delta M_1 - \Delta M_2 = \rho_1 A_1 v_1 \Delta t - \rho_2 A_2 v_2 \Delta t.$$

Conservation of energy is applied in a similar manner: It is assumed that the change in energy of the volume of the streamtube bounded by A_1 and A_2 is due entirely to energy entering or leaving through one or the other of these two boundaries. Clearly, in a more complicated situation such as a fluid flow coupled with radiation, such conditions are not met. Nevertheless, assuming this to be the case and assuming the flow is steady so that the net change in the energy is zero,

$$0 = \Delta E_1 - \Delta E_2$$

where ΔE_1 and ΔE_2 are the energy entering through A_1 and leaving through A_2, respectively. The energy entering through A_1 is the sum of the kinetic energy entering, the energy entering in the form of potential gravitational energy of the fluid, the fluid thermodynamic internal energy per unit of mass (ε_1) entering, and the energy entering in the form of mechanical $p\,dV$ work:

$$\Delta E_1 = \left(\tfrac{1}{2}\rho_1 v_1^2 + \Psi_1 \rho_1 + \epsilon_1 \rho_1 + p_1\right) A_1 v_1 \Delta t$$

where $\Psi = gz$ is a force potential due to the Earth's gravity, g is acceleration due to gravity, and z is elevation above a reference plane. A similar expression for ΔE_2 may easily be constructed. So now setting $0 = \Delta E_1 \, \Delta E_2$:

$$0 = \left(\tfrac{1}{2}\rho_1 v_1^2 + \Psi_1\rho_1 + \epsilon_1\rho_1 + p_1\right) A_1 v_1 \Delta t - \left(\tfrac{1}{2}\rho_2 v_2^2 + \Psi_2\rho_2 + \epsilon_2\rho_2 + p_2\right) A_2 v_2 \Delta t$$

which can be rewritten as:

$$0 = \left(\tfrac{1}{2}v_1^2 + \Psi_1 + \epsilon_1 + \frac{p_1}{\rho_1}\right)\rho_1 A_1 v_1 \Delta t - \left(\tfrac{1}{2}v_2^2 + \Psi_2 + \epsilon_2 + \frac{p_2}{\rho_2}\right)\rho_2 A_2 v_2 \Delta t$$

Now, using the previously-obtained result from conservation of mass, this may be simplified to obtain,

$$\frac{1}{2}v^2 + \varnothing + \dot{o} + \frac{p}{\rho} = \text{constant} \equiv b$$

which is the Bernoulli equation for compressible flow.

An equivalent expression can be written in terms of fluid enthalpy (h):

$$\tfrac{1}{2}v^2 + \Psi + h = \text{constant} \equiv b$$

Applications

Condensation visible over the upper surface of an Airbus A340 wing caused by the fall in temperature accompanying the fall in pressure.

In modern everyday life there are many observations that can be successfully explained by application of Bernoulli's principle, even though no real fluid is entirely inviscid and a small viscosity often has a large effect on the flow.

- Bernoulli's principle can be used to calculate the lift force on an airfoil, if the behaviour of the fluid flow in the vicinity of the foil is known. For example, if the air flowing past the top surface of an aircraft wing is moving faster than the air flowing past the bottom surface, then Bernoulli's principle implies that the pressure on the surfaces of the wing will be lower above than below. This pressure difference results in an upwards lifting force. Whenever the distribution of speed past the top and bottom surfaces of a wing is known, the lift forces can be calculated (to a good approximation) using Bernoulli's equations – established by Bernoulli over a century before the first man-made wings were used for the purpose of flight. Bernoulli's principle does not explain why the air flows faster past the top of the wing and slower past the underside.

- The carburettor used in many reciprocating engines contains a venturi to create a region of low pressure to draw fuel into the carburettor and mix it thoroughly with the incoming air. The low pressure in the throat of a venturi can be explained by Bernoulli's principle; in the narrow throat, the air is moving at its fastest speed and therefore it is at its lowest pressure.
- An injector on a steam locomotive (or static boiler).
- The pitot tube and static port on an aircraft are used to determine the airspeed of the aircraft. These two devices are connected to the airspeed indicator, which determines the dynamic pressure of the airflow past the aircraft. Dynamic pressure is the difference between stagnation pressure and static pressure. Bernoulli's principle is used to calibrate the airspeed indicator so that it displays the indicated airspeed appropriate to the dynamic pressure.
- A De Laval nozzle utilizes Bernoulli's principle to create a force by turning pressure energy generated by the combustion of propellants into velocity. This then generates thrust by way of Newton's third law of motion.
- The flow speed of a fluid can be measured using a device such as a Venturi meter or an orifice plate, which can be placed into a pipeline to reduce the diameter of the flow. For a horizontal device, the continuity equation shows that for an incompressible fluid, the reduction in diameter will cause an increase in the fluid flow speed. Subsequently, Bernoulli's principle then shows that there must be a decrease in the pressure in the reduced diameter region. This phenomenon is known as the Venturi effect.
- The maximum possible drain rate for a tank with a hole or tap at the base can be calculated directly from Bernoulli's equation, and is found to be proportional to the square root of the height of the fluid in the tank. This is Torricelli's law, showing that Torricelli's law is compatible with Bernoulli's principle. Viscosity lowers this drain rate. This is reflected in the discharge coefficient, which is a function of the Reynolds number and the shape of the orifice.
- The Bernoulli grip relies on this principle to create a non-contact adhesive force between a surface and the gripper.
- Bernoulli's principle is also applicable in the swinging of a cricket ball. During a cricket match, bowlers continually polish one side of the ball. After some time, one side is quite rough and the other is still smooth. Hence, when the ball is bowled and passes through air, the speed on one side of the ball is faster than on the other, due to this difference in smoothness, and this results in a pressure difference between the sides; this leads to the ball rotating ("swinging") while travelling through the air, giving advantage to the bowlers.

Misunderstandings about the Generation of Lift

Many explanations for the generation of lift (on airfoils, propeller blades, etc.) can be found; some of these explanations can be misleading, and some are false. There has been debate about whether lift is best introduced to students using Bernoulli's principle or Newton's laws of motion. Modern writings agree that both Bernoulli's principle and Newton's laws are relevant, and either can be used to correctly describe lift.

Several of these explanations use the Bernoulli principle to connect the flow kinematics to the flow-induced pressures. In cases of incorrect (or partially correct) explanations relying on the Bernoulli principle, the errors generally occur in the assumptions on the flow kinematics and how these are produced. It is not the Bernoulli principle itself that is questioned, because this principle is well established (the airflow above the wing *is* faster, the question is *why* it is faster).

Misapplications of Bernoulli's Principle in Common Classroom Demonstrations

There are several common classroom demonstrations that are sometimes incorrectly explained using Bernoulli's principle. One involves holding a piece of paper horizontally so that it droops downward and then blowing over the top of it. As the demonstrator blows over the paper, the paper rises. It is then asserted that this is because "faster moving air has lower pressure".

One problem with this explanation can be seen by blowing along the bottom of the paper: were the deflection due simply to faster moving air one would expect the paper to deflect downward, but the paper deflects upward regardless of whether the faster moving air is on the top or the bottom. Another problem is that when the air leaves the demonstrator's mouth it has the *same* pressure as the surrounding air; the air does not have lower pressure just because it is moving; in the demonstration, the static pressure of the air leaving the demonstrator's mouth is *equal* to the pressure of the surrounding air. A third problem is that it is false to make a connection between the flow on the two sides of the paper using Bernoulli's equation since the air above and below are *different* flow fields and Bernoulli's principle only applies within a flow field.

As the wording of the principle can change its implications, stating the principle correctly is important. What Bernoulli's principle actually says is that within a flow of constant energy, when fluid flows through a region of lower pressure it speeds up and vice versa. Thus, Bernoulli's principle concerns itself with *changes* in speed and *changes* in pressure *within* a flow field. It cannot be used to compare different flow fields.

A correct explanation of why the paper rises would observe that the plume follows the curve of the paper and that a curved streamline will develop a pressure gradient perpendicular to the direction of flow, with the lower pressure on the inside of the curve. Bernoulli's principle predicts that the decrease in pressure is associated with an increase in speed, i.e. that as the air passes over the paper it speeds up and moves faster than it was moving when it left the demonstrator's mouth. But this is not apparent from the demonstration.

Other common classroom demonstrations, such as blowing between two suspended spheres, inflating a large bag, or suspending a ball in an airstream are sometimes explained in a similarly misleading manner by saying "faster moving air has lower pressure".

Open-channel Flow

Open-channel flow, a branch of hydraulics and fluid mechanics, is a type of liquid flow within a conduit with a free surface, known as a channel. The other type of flow within a conduit is pipe flow. These two types of flow are similar in many ways but differ in one important respect: the free surface. Open-channel flow has a free surface, whereas pipe flow does not.

Classifications of Flow

Open-channel flow can be classified and described in various ways based on the change in flow depth with respect to time and space. The fundamental types of flow dealt with in open-channel hydraulics are:

Time as the Criterion

- Steady flow: The depth of flow does not change over time, or if it can be assumed to be constant during the time interval under consideration.
- Unsteady flow: The depth of flow does change with time.

Space as the Criterion

- Uniform flow: The depth of flow is the same at every section of the channel. Uniform flow can be steady or unsteady, depending on whether or not the depth changes with time, (although unsteady uniform flow is rare).
- Varied flow: The depth of flow changes along the length of the channel. Varied flow technically may be either steady or unsteady.
- Varied flow can be further classified as either rapidly or gradually varied.
- Rapidly varied flow: The depth changes abruptly over a comparatively short distance. Rapidly varied flow is known as a local phenomenon. Examples are the hydraulic jump and the hydraulic drop.
- Gradually varied flow: The depth changes over a long distance.
- Continuous flow: The discharge is constant throughout the reach of the channel under consideration. This is often the case with a steady flow. This flow is considered continuous and therefore can be described using the continuity equation for continuous steady flow.
- Spatially varied or discontinuous flow: The discharge of a steady flow is non-uniform along a channel. This happens when water enters and/or leaves the channel along the course of flow. An example of flow entering a channel would be a road side gutter. An example of flow leaving a channel would be an irrigation channel. This flow can be described using the continuity equation for continuous unsteady flow requires the consideration of the time effect and includes a time element as a variable.

States of Flow

The behavior of open channel flow is governed by the effects of viscosity and gravity relative to the inertial forces of the flow. Surface tension has a minor contribution but does not play a significant enough role in most circumstances to be a governing factor. Depending on the effect of viscosity relative to inertia, as represented by the Reynolds number, the flow can be either laminar, turbulent, or transitional.

Pipe Flow

Pipe flow, a branch of hydraulics and fluid mechanics, is a type of liquid flow within a closed conduit (conduit in the sense of a means of containment). The other type of flow within a conduit is open channel flow. These two types of flow are similar in many ways, but differ in one important aspect. Pipe flow does not have a free surface which is found in open-channel flow. Pipe flow, being confined within closed conduit, does not exert direct atmospheric pressure, but does exert hydraulic pressure on the conduit.

Not all flow within a closed conduit is considered pipe flow. Storm sewers are closed conduits but usually maintain a free surface and therefore are considered open-channel flow. The exception to this is when a storm sewer operates at full capacity, and then can become pipe flow.

Energy in pipe flow is expressed as head and is defined by the Bernoulli equation. In order to conceptualize head along the course of flow within a pipe, diagrams often contain a hydraulic grade line (HGL). Pipe flow is subject to frictional losses as defined by the Darcy-Weisbach formula.

States of Flow

The behavior of pipe flow is governed mainly by the effects of viscosity and gravity relative to the inertial forces of the flow. Depending on the effect of viscosity relative to inertia, as represented by the Reynolds number, the flow can be either laminar or turbulent. At a Reynolds number below the critical value of approximately 2040 pipe flow will ultimately be laminar, whereas above the critical value turbulent flow can persist. In addition, the transition between laminar flow and turbulence can be sensitive to disturbance levels and imperfections.

Flow through pipes can roughly be divided into two:

- Laminar flow
- Turbulent flow

General Characteristics of Pipe Flow

The general method of transporting fluid (liquid or gas) is the flow through a closed conduit. It is commonly called 'pipe' if it is of round cross-section and 'duct' if the cross-section is not round. The common examples include water pipes, hydraulic hoses, air distribution in a duct in an air conditioning plant etc. The main driving force for the flow to occur is the pressure differential at both ends of the pipe and the walls of the pipe which is designed to withstand this pressure difference without any undue distortion of the shape.

The basic governing equations such as mass, momentum and energy conservation can be applied for viscous, incompressible fluids in pipes and ducts. Following assumptions are made for present analysis;

Cross-section of the duct is circular unless otherwise specified.

The pipe is completely filled with fluid being transported otherwise the flow may be treated as open-channel flow where gravity alone is the driving force.

The driving potential is the pressure difference across the pipe.

The flow of fluid in a pipe may be laminar, turbulent or transitional depending upon the flow rates. Such a flow in the pipe is characterized by a dimensionless number called "Reynolds number" and is defined as,

$$R_e = \frac{\rho V D}{\mu}$$

where ρ and μ are the density and viscosity of the flowing fluid, D is the diameter of the pipe and V is the average velocity in the pipe.

Pressure Surge in Pipes

Surge pressures, or pressure surges, occur in a pipe transporting fluid as a result of a change in the flow velocity, e. g. if a valve is closed or opened too rapidly. If a valve is closed rapidly, the energy of the fluid moving forward in the pipe leads to a sudden rise in pressure due to compression upstream of the valve.

Negative pressure will develop immediately downstream of the valve as a result of the fluid being still in motion, causing a temporary separation of the liquid column followed by a reverse flow back towards the valve if the pressure gradient is high enough. This may lead to the destruction of both the valve and the pipe.

The highest surge pressures often occur when the maximum number of pumps in operation at a pumping station stop as a result of a power failure. As the change in pressure and velocity is not limited to the point of disturbance, it continues up- and downstream at pressure wave propagation velocity.

The boundary conditions determine the degree to which waves are reflected at locations of non-steady flow (e. g. pipe branches, valves, changes in cross-sections, tanks) resulting in a phase or amplitude reversal.

The condition observed at a defined location and time is determined by the superposition of all waves arriving at this defined location at the defined time. The pressure fluctuations and resultant maximum pressures cause extremely high loads for the system.

Although the pressure level cannot fall under that of vapour pressure, the minimum pressure may reach that level. If this situation persists for a prolonged period of time, a cavitation zone will develop leading to the separation of the liquid column into two columns with resultant flow separation. Following their change in velocity direction, the liquid columns will reverse and often collide with high velocity differentials, causing a new pressure surge.

This pressure surge is often significantly greater than the original surge.

Maximum Permissible Negative Pressures Depending on Piping Material and Layout

- Max. 0.2 bar for plastic and fibre-glass reinforced plastic pipes.
- Approx. 0.4 bar abs. (corresponding to a negative pressure of 0.6 bar) for welded steel pipes (depending on the wall thickness) for connection to steel or cast pipes via sockets.

The minimum pressure should never fall below the atmospheric pressure at any point in the system which means that this system pressure must also be maintained following a power supply failure. The systems most at risk from pressure surges are low- and medium-pressure systems and not high-pressure systems.

According to Joukowsky, the following applies to the maximum change in pressure:

$$\Delta p = \rho \cdot a \cdot \Delta v$$

Δp Change in pressure in bar

ρ Density of water in kg/m^3

a Pressure wave propagation velocity in m/s

Δv Change in flow velocity in m/s

Für die maximale Änderung der Druckhöhe gilt:

$$\Delta H_{max} = \frac{a}{g} \cdot v_0$$

ΔH_{max} Maximum change in pressure head in m

g Acceleration due to gravity in m/s^2

v_0 Undisturbed velocity in m/s

The maximum change in pressure head (ΔH_{max}) may only develop if this applies to the time (t_s) during which the full change in velocity (v_0) takes place:

$$t_s \gg t_r = 2.\frac{L}{a}$$

t_r Reflection time L Distance between the nearest reflection point and the point of disturbance a Pressure wave propagation velocity

According to Joukowsky, the pressure change is superimposed both positively and negatively by the steady-state pressure upstream of the point of disturbance. This may lead to negative pressures in the piping.

If the time (t_s) for the change in velocity is longer than the reflection time (t_r), then the change in velocity within the reflection time must be used to calculate the maximum change in pressure head (ΔH) at v_o.

The relationship between the switching and reflection time implicitly expresses that the longer the switching time (t_s) is in relation to the system's reflection time (t_r), the smaller the changes in pressure (Δp) become. The ratio should be around 5 to 1 in order to keep the consequences of a pressure surge within permissible limits.

A differentiation is made between two cases; the measures taken to limit the maximum pressure will have to become effective the second step only:

t_s is given and t_r must be reduced in accordance with the ratio above by:

- Keeping the piping as short as possible.
- Providing intermediate reflection points for long pipes (e.g. provision of surge tanks or drums at the highest point, stand pipes at intermediate and highest points, venting the pipe at the highest points, provision of a one-way surge tank).

t_r is given and ts must be increased in accordance with the ratio above by:

- Selecting appropriate valve closure characteristics.
- Extending the pump's run-down time after it is switched off (e.g. via flywheel mass).
- Providing further liquid from a tank (e. g. accumulator, stand pipe, one-way surge tank).
- Discharging suddenly accumulated masses of fluid (e.g. via additional outlets, safety valves, bypass).

According to Joukowsky's equation, the change in pressure is low if the changes in the velocity of the fluid flow (Δv) and pressure wave propagation (a) remain low. This applies to light fluids, low flow velocities and low sound velocities (e. g. via large-diameter piping).

If the problems resulting from surge pressure have not already been considered at the planning stage, the possibilities of influencing these parameters are often very low or involve considerable effort.

Impermissible high or low pressures can often only be avoided by employing appropriate surge control systems. The selection and sizing of required surge control systems are so complex that computerised methods are needed to calculate them.

References

- Falkovich, Gregory (2011). Fluid Mechanics (A short course for physicists). Cambridge University Press. ISBN 978-1-107-00575-4
- Avila, K.; D. Moxey; A. de Lozar; M. Avila; D. Barkley; B. Hof (July 2011). "The Onset of Turbulence in Pipe Flow". Science. 333 (6039): 192–196. Bibcode:2011Sci...333..192A. doi:10.1126/science.1203223. PMID 21737736
- Battjes, J. A.; Labeur, R. J. (2017). Unsteady flow in open channels. Cambridge University Press. Doi:10.1017/9781316576878. ISBN 978-1-107-15029-4
- Darrigol, O.; Frisch, U. (2008), "From Newton›s mechanics to Euler›s equations", Physica D: Nonlinear Phenomena, 237 (14–17): 1855–1869, Bibcode:2008PhyD..237.1855D, doi:10.1016/j.physd.2007.08.003
- Mulley, Raymond (2004). Flow of Industrial Fluids: Theory and Equations. CRC Press. Pp. 43–44. ISBN 978-0-8493-2767-4
- Avila, K.; D. Moxey; A. De Lozar; M. Avila; D. Barkley; B. Hof (July 2011). "The Onset of Turbulence in Pipe Flow". Science. 333 (6039): 192–196. Bibcode:2011Sci...333..192A. Doi:10.1126/science.1203223. PMID 21737736
- Surge-pressure, centrifugal-pump-lexicon: ksb.com, Retrieved 23 February, 2019
- Kundu, Pijush K.; Cohen, Ira M. (2008). Fluid Mechanics (4th rev. Ed.). Academic Press. ISBN 978-0-12-373735-9

Permissions

All chapters in this book are published with permission under the Creative Commons Attribution Share Alike License or equivalent. Every chapter published in this book has been scrutinized by our experts. Their significance has been extensively debated. The topics covered herein carry significant information for a comprehensive understanding. They may even be implemented as practical applications or may be referred to as a beginning point for further studies.

We would like to thank the editorial team for lending their expertise to make the book truly unique. They have played a crucial role in the development of this book. Without their invaluable contributions this book wouldn't have been possible. They have made vital efforts to compile up to date information on the varied aspects of this subject to make this book a valuable addition to the collection of many professionals and students.

This book was conceptualized with the vision of imparting up-to-date and integrated information in this field. To ensure the same, a matchless editorial board was set up. Every individual on the board went through rigorous rounds of assessment to prove their worth. After which they invested a large part of their time researching and compiling the most relevant data for our readers.

The editorial board has been involved in producing this book since its inception. They have spent rigorous hours researching and exploring the diverse topics which have resulted in the successful publishing of this book. They have passed on their knowledge of decades through this book. To expedite this challenging task, the publisher supported the team at every step. A small team of assistant editors was also appointed to further simplify the editing procedure and attain best results for the readers.

Apart from the editorial board, the designing team has also invested a significant amount of their time in understanding the subject and creating the most relevant covers. They scrutinized every image to scout for the most suitable representation of the subject and create an appropriate cover for the book.

The publishing team has been an ardent support to the editorial, designing and production team. Their endless efforts to recruit the best for this project, has resulted in the accomplishment of this book. They are a veteran in the field of academics and their pool of knowledge is as vast as their experience in printing. Their expertise and guidance has proved useful at every step. Their uncompromising quality standards have made this book an exceptional effort. Their encouragement from time to time has been an inspiration for everyone.

The publisher and the editorial board hope that this book will prove to be a valuable piece of knowledge for students, practitioners and scholars across the globe.

Index